The Challenges of Energy Harvesting

Edited by Sadia Ameen,
M. Shaheer Akhtar and Ing Kong

Published in London, United Kingdom

The Challenges of Energy Harvesting
http://dx.doi.org/10.5772/intechopen.111075
Edited by Sadia Ameen, M. Shaheer Akhtar and Ing Kong

Contributors
Abdenasser Lamk, Ahmed El Yousfi, Anna Montini, Chiang Liang Kok, Chiang Liang Kok, Daniel Segovia Vargas, em, Furqan Ali Akhtar, Guangming Ren, Haneea Isaad, Hongyi Goh, Jovan Heng, Mahmood Ahmad, Marcos Katz, Mariella Särestöniemi, Mohsen Hossein Zadeh, Moslem Forouzesh, Rui Alexandre Castanho, Syifaul Fuada, Tee Hui Teo, Tee Hui Teo, Xianlong Zhang, Xinlong Zhang, Ziyafat Guluzade, Zyon Tham

First published in London, United Kingdom, 2025 by IntechOpen
IntechOpen is the global imprint of INTECHOPEN LIMITED, registered in England and Wales, registration number: 11086078, 167-169 Great Portland Street, London, W1W 5PF, United Kingdom

For EU product safety concerns: IN TECH d.o.o., Prolaz Marije Krucifikse Kozulić 3, 51000 Rijeka, Croatia, info@intechopen.com or visit our website at intechopen.com.

British Library Cataloguing-in-Publication Data
A catalogue record for this book is available from the British Library

The Challenges of Energy Harvesting
Edited by Sadia Ameen, M. Shaheer Akhtar and Ing Kong
p. cm.
Print ISBN 978-1-83634-386-8
Online ISBN 978-1-83634-385-1
eBook (PDF) ISBN 978-1-83634-387-5

If disposing of this product, please recycle the paper responsibly.

Meet the editors

Professor Sadia Ameen is working as an Associate Professor in the Department of Bio-Convergence Science at the Jeongeup Campus, Jeonbuk National University, and as an Adjunct Professor in the Department of Bioactive Material Sciences at Jeonbuk National University in Jeonju, Republic of Korea. Her current research focuses on solar cells, perovskite solar cells, organic solar cells, sensors, catalysts, and other optoelectronic devices. She specializes in the synthesis and application of clean energy materials. She is the recipient of a Gold Medal for academic achievement and several scientific honors, including the Excellence in Research Award, Outstanding Scientist Award, Asia's Top-50 Scientist Award, Certificate of Excellence, Best Researcher Award, and selected in the Stanford University list of World's Top 2% scientists from 2019-2024. She has authored or co-authored more than 170 peer-reviewed papers in the fields of solar cells, catalysts, and sensors, as well as contributed to book chapters and edited several books.

Prof. Dr. M. Shaheer Akhtar completed his Ph.D. in Chemical Engineering in 2008 at Jeonbuk National University, Republic of Korea. Presently, he is working as a Full Professor at Jeonbuk National University, Republic of Korea. His research interests constitute the synthesis, fabrication, and characterization of energy materials, composite materials, polymer-based solid-state films, solid polymer electrolytes, and electrode materials for hybrid organic-inorganic solar cells, perovskite solar cells, supercapacitors, small-molecule-based organic solar cells, sensing technology, environmental engineering, and sustainability.

Ing Kong is an Associate Professor in the Department of Engineering at La Trobe University. She is the Leader of the Advanced Polymer and Composites Group on the Bendigo Campus. She has established her research expertise in additive manufacturing technology and the development of advanced materials, particularly functional polymer composites. Across her career, she has authored 114 publications and secured over $6 M of external grants. She has promoted interdisciplinary and joint research among the research groups from national and international universities as well as industry partners, including Ceramic Oxide Fabricators, Ritchie Technology, MoreDent, etc. She serves as an editorial board member for several international scientific journals and a peer reviewer for many materials and composites journals.

Contents

Preface

The growing demand for sustainable and efficient energy solutions has emerged as one of the defining issues. As the world's population grows and our reliance on electronic devices, smart systems, and renewable technologies grows, the demand for reliable, self-sufficient, and environmentally friendly energy sources is also increasing. One of the emerging ideas is energy harvesting, which involves capturing and storing energy from natural sources such as light, heat, motion, and radio frequency. While the concept has transformative potential, there are several challenges, limitations, and unanswered issues. The book *The Challenges of Energy Harvesting* aims to investigate these issues.

This book is written as an academic resource for engineers, researchers, graduate students, and policymakers who are interested in sustainable energy innovation. *The Challenges of Energy Harvesting* will serve as a practical guide for understanding the issues of energy harvesting. The book's chapters are organized around specific subjects, such as problems and future trends in wireless power transfer for biomedical implants, developments in wireless charging, energy harvesting in the smart grid context, etc. The book tackles challenges such as energy inconsistency, conversion inefficiencies, cost constraints, scale issues, and material degradation.

We believe that this book will assist readers in understanding the benefits and risks of energy harvesting. Whether used as a reference, a teaching tool, or a starting point for new research, the book is intended to discuss the most critical technological frontiers. We want to thank the authors, reviewers, and colleagues who worked on this book. We would not have been able to publish this book without their help. We also thank IntechOpen for their help and encouragement during the preparation process.

Sadia Ameen
Department of Bio-Convergence Science,
Jeonbuk National University,
Jeongeup Campus, Republic of Korea

Department of Bioactive Material Sciences,
Jeonbuk National University,
Jeonju, Republic of Korea

M. Shaheer Akhtar
Graduate School of Integrated Energy-AI,
Jeonbuk National University,
Jeonju, Republic of Korea

Department of Engineering,
School of Computing, Engineering and Mathematical Sciences,
La Trobe University,
Bendigo, Australia

New & Renewable Energy Material Development Center (NewREC),
Jeonbuk National University,
Jeonbuk, Republic of Korea

Ing Kong
Department of Engineering,
School of Computing, Engineering and Mathematical Sciences,
La Trobe University,
Bendigo, Australia

Chapter 1

Perspective Chapter: Wearable Energy Harvesting for Personal Navigation

Tee Hui Teo, Chiang Liang Kok, Xianlong Zhang and Jovan Heng

Abstract

Energy harvesting technologies are driving sustainable innovations across various domains, from wearable devices to advanced satellite positioning systems. This chapter highlights transformative projects showcasing the potential of energy harvesting. The energy harvesting combat boot exemplifies the ability to convert kinetic energy from human motion into electrical power, enabling operations such as satellite positioning without external batteries. Advanced wearable devices, including global positioning system (GPS)-integrated systems, illustrate how energy harvesting supports efficient navigation and tracking. Emerging methods like triboelectric and piezoelectric harvesters, alongside cutting-edge materials such as carbon nanotubes, demonstrate scalable solutions for low-power applications. Innovations in satellite energy systems, such as thermoelectric generators and hybrid designs, further extend the versatility of energy harvesting by converting ambient energy into usable power for communication and navigation. Despite challenges like intermittent energy sources and integration complexities, these advancements underscore the critical role of energy harvesting in achieving autonomy, efficiency, and environmental sustainability.

Keywords: energy harvesting, wearable devices, battery management systems, satellite positioning systems, low-power electronics

1. Introduction

Energy harvesting technologies are revolutionizing sustainable energy practices across various fields [1], particularly in wearable devices where human motion can be transformed into usable electrical energy. Among these innovations, the energy harvesting combat boot exemplifies the potential of wearable systems to deliver practical solutions. This technology utilizes air bulbs embedded in the sole of the boot, which compress during walking to drive micro-turbines connected to small direct current (DC) motors, producing electrical power. This harvested energy powers a passive GPS module, enabling satellite positioning and communication without reliance on external batteries. By leveraging the natural motion of walking, the combat boot provides an efficient and sustainable power source for use in remote or resource-constrained environments. With its ability to generate power continuously, this innovation offers critical applications for military personnel, outdoor enthusiasts, and emergency responders. This chapter explores the design and functionality of the energy harvesting combat boot, highlights its comparative performance against similar technologies, and discusses potential improvements to enhance its efficiency and user experience. Ultimately, this wearable energy harvesting system underscores the transformative role of such technologies in promoting sustainability and energy independence.

2. Design and functionality of the energy harvesting combat boot

The energy harvesting combat boot is a sophisticated wearable technology designed to transform the mechanical energy of human motion into usable electrical energy [1]. **Figure 1** displays a schematic of the energy harvesting system embedded in the sole of the boot. Central to its functionality are air bulbs embedded within the sole, strategically positioned in high-pressure regions such as the heel and ball areas. These air bulbs compress with each foot strike, producing airflow that drives micro-turbines. The turbines are designed to operate bi-directionally, ensuring consistent power generation regardless of airflow direction. The system's airflow mechanism is integral to its performance. Each step compresses the air bulbs, forcing air through the narrow channels that power micro-turbines. These turbines are connected to DC motors, which convert mechanical energy into electrical energy. The generated power is then stored in a supercapacitor, which serves as an intermediary storage device before charging the battery. At a walking speed of 3.0 mph (1 Hz footstep frequency), the prototype boot delivered an average continuous power output of 86.4 mW and a peak power of 516 mW. During high-impact movements like lunges, the system achieved a peak power of 679 mW and an average power of 106 mW. These metrics demonstrate the system's ability to harness varying energy levels depending on movement intensity. The energy harvesting system captured approximately 13.4% of the theoretical maximum foot strike energy. While this represents a significant milestone, it also highlights the challenge of improving energy conversion efficiency without compromising user comfort. Durability and wearability remain crucial considerations. The additional weight of the system, estimated at 130 g (a 17% weight increase), was optimized to minimize ergonomic impact. Future designs

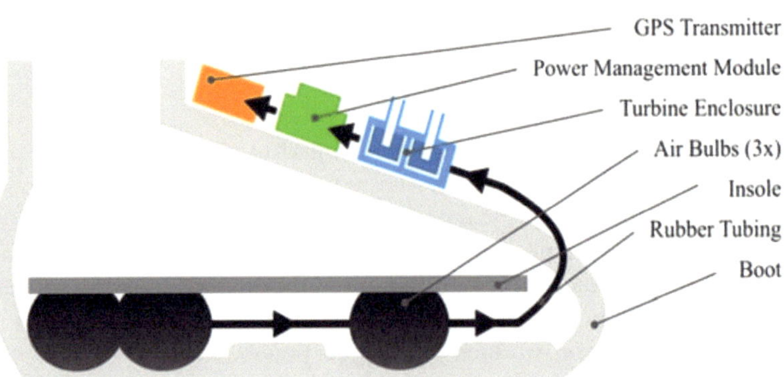

Figure 1.
Schematic of energy harvesting system [1].

aim to reduce this weight to 20 g through advancements in materials and component miniaturization. The air bulb-turbine system's efficiency compares favorably to piezoelectric-based energy harvesters embedded in footwear, which typically achieve lower-energy outputs. Piezoelectric systems rely on compressive stress to generate electricity but often face challenges related to structural integrity and energy density. In contrast, the air-driven system demonstrated durability and scalability, making it suitable for diverse applications, including military and outdoor scenarios. While the system performed well under standard testing conditions, its efficiency could be further enhanced by optimizing airflow dynamics. Computational fluid dynamics models may provide valuable insights into turbine blade designs that maximize power generation. Additionally, incorporating hybrid systems that combine kinetic and solar energy harvesting could address limitations in low-motion scenarios. The materials and placement of the air bulbs remain critical to the system's performance. Lightweight and durable components ensure minimal impact on user comfort, while precise positioning maximizes energy output. Furthermore, the modular design allows for integration into various types of footwear, expanding its potential applications. Despite these advancements, maintaining system durability under prolonged use and extreme environmental conditions remains an ongoing challenge.

3. Other forms of energy harvesting

Other forms of energy harvesting are available, providing diverse solutions for wearable and portable applications beyond the air bulb-turbine system. These technologies harness kinetic, thermal, and environmental energy, enabling sustainable power solutions for low-power devices. Kinetic energy harvesting [2] is one of the most promising approaches for wearable applications. It involves converting mechanical movements, such as human motion, into electrical energy using piezoelectric, triboelectric, or electromagnetic systems. Piezoelectric energy harvesting systems embedded in footwear effectively capture energy from footfalls. PZT piezoelectric elements have been shown to generate sufficient power to charge supercapacitors while maintaining wearer comfort and mobility [3]. Biomechanical knee-joint energy harvesters convert bidirectional knee motion into unidirectional rotation, driving an electromagnetic generator. These systems achieve an average power output of 3.6 W at a walking speed of 1.5 m/s, sufficient for powering portable electronic devices [4]. Triboelectric nanogenerators (TENGs) integrated into textiles demonstrate practical applications for wearable systems. These devices produce an energy output of 93 mW/m^2 and are compatible with washable fabrics, making them suitable for daily use [5]. Carbon nanotube yarn harvesters are another innovative method for converting kinetic energy from mechanical motion into electrical energy by using the torsional and tensile properties of nanotube yarns to achieve exceptional energy density and efficiency. By integrating this technology into textiles or small devices, energy can be harvested seamlessly during daily activities. Recent advancements in this field have shown these systems capable of producing up to 250 W/kg of peak electrical power when subjected to mechanical stretching [6]. Heel-strike energy harvesters using dielectric elastomers shown in **Figure 2** have achieved outputs of 0.8 Joules per step with an efficiency of 33%. These designs are particularly effective for scalable and lightweight footwear-based energy harvesting solutions [7]. Polyvinylidene fluoride (PVDF)-based nanogenerators have advanced piezoelectric energy harvesting further. These systems demonstrate power densities of 0.3 $\mu W/cm^2$ at 200 Hz, with

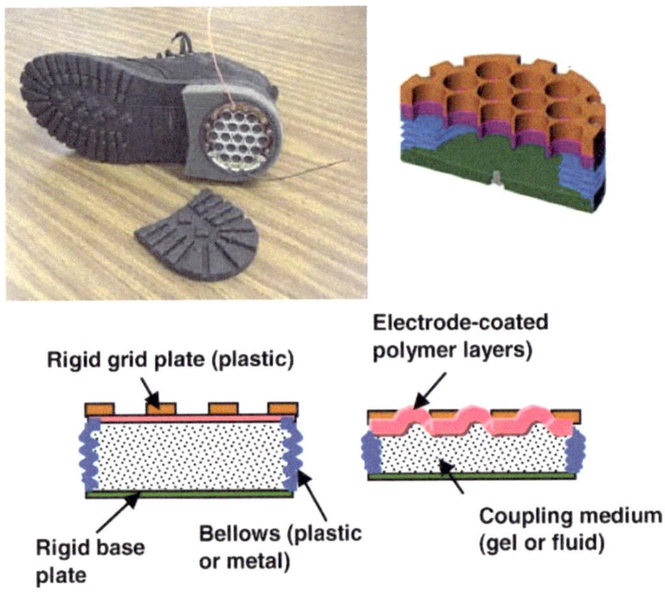

Figure 2.
Heel-strike generator based on a dielectric elastomer [7].

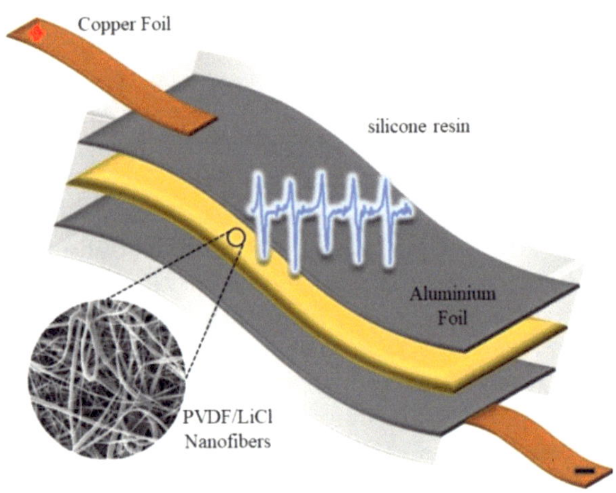

Figure 3.
Schematic structure of PVDF power-generating device [8].

flexible and lightweight designs that are well-suited for integration into wearable fabrics, with **Figure 3** showing the schematic structure of the PVDF nanofibers in creating a nanofibrous power generator [8].

Thermoelectric generators (TEGs) utilize temperature gradients between the skin and the environment to produce electricity. These systems have been successfully integrated into wearable fabrics, providing continuous energy for low-power electronics. They are particularly effective in healthcare applications, where body heat is a consistent energy source [9]. Photovoltaic cells have also emerged as a viable option for wearable devices, particularly when combined with flexible substrates. Advances in 2D materials have enabled the development of lightweight solar cells with enhanced efficiency, allowing them to be integrated into wearable systems for outdoor energy harvesting [10]. Emerging materials also offer exciting opportunities for energy harvesting. Hybrid systems that combine multiple energy harvesting methods, such as kinetic and solar, are gaining traction. These systems provide consistent energy output by compensating for the limitations of individual methods. For instance, a hybrid nanogenerator combining triboelectric and piezoelectric elements has demonstrated the ability to power wearable medical sensors and low-power IoT devices [11].

4. Energy harvesting for satellite positioning

The integration of energy harvesting systems with satellite positioning technology represents a transformative approach for portable and wearable devices. Satellite positioning systems, such as GPS, are essential for navigation and tracking across numerous applications, but their reliance on external power sources can be a limiting factor, particularly in remote or resource-constrained environments. Energy harvesting provides a sustainable solution to this challenge by enabling devices to operate independently of traditional batteries. Recent advancements have demonstrated significant progress in reducing power consumption for wearable positioning systems while maintaining accuracy. For instance, a compact wearable positioning system utilizing ultrasonic transmissions and particle filters requires less than 10 mW to operate, highlighting its potential for low-power satellite positioning integration in wearable devices [12]. Similarly, GPS-enabled smartwatches and wearable devices now offer varying levels of measurement precision based on environmental conditions. A comparative study of GPS performance in smartwatches demonstrated how factors like signal obstruction and device design influence accuracy. For example, open-sky conditions allowed precise location tracking, whereas environments with multipath effects, such as urban canyons, introduced notable inaccuracies [13]. These findings emphasize the need for energy-efficient geo-localization solutions tailored for wearable technology. Hybrid positioning systems, which combine GPS with Wi-Fi signals, have also emerged as a viable solution. A wearable camera system demonstrated how this approach minimizes energy consumption while maintaining positional data accuracy. The system achieved a mean localization error of 27 m while avoiding over 80% of unnecessary position estimates by leveraging accelerometer data for activity detection [14]. Advanced self-powered systems represent the frontier of sustainable positioning technology. For instance, a self-powered position monitoring system integrates an insole-type triboelectric nanogenerator (IW-TENG) with Bluetooth beacons to harvest human kinetic energy, shown in **Figure 4**. This system transmits location data every 3–4 steps, with a single Bluetooth transmission consuming only 136 μJ and achieving a range of 100 m [15].

Microwave energy harvesting for satellite applications has also been explored. A system using K-band microwave energy has shown feasibility for powering autonomous wireless sensors on geostationary satellites, leveraging the existing electromagnetic environment for energy generation [16]. TEGs have shown great promise for satellites by converting waste heat into electrical power. These systems offer maintenance-free, vibration-free operation, making them ideal for harsh space environments. TEGs have the potential to increase satellite energy efficiency by utilizing heat flows that are typically radiated into space, providing an independent power source alongside solar panels [17]. Optimized thermoelectric devices also enhance energy harvesting capabilities for low-temperature gradient applications. Advanced

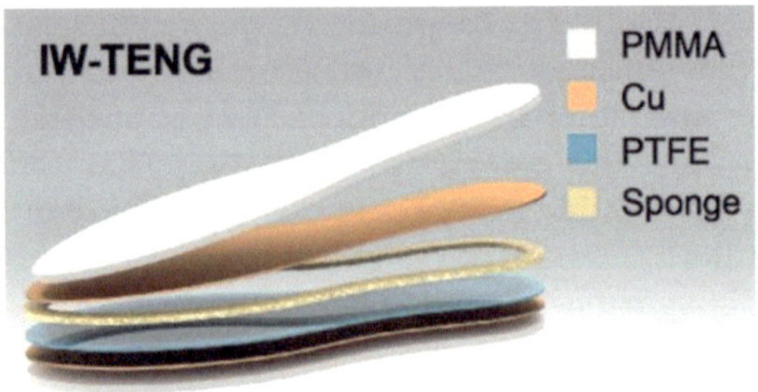

Figure 4.
Exploded diagram of the IW-TENG in a Shoe Insole [15].

designs employing thin-film thermoelectric modules have shown improved performance in aerospace environments, where traditional bulk thermoelectric devices struggle with efficiency due to mismatched thermal resistances. These advancements highlight the importance of tailoring thermoelectric designs to the specific environmental and application requirements [18]. A metamaterial-based energy harvesting structure capable of supporting multiband applications has also been developed. This system can harvest energy from frequencies used in GSM, satellite communication, and ISM bands, making it a versatile option for wearable and satellite communication devices [19]. Finally, innovations such as the Battery-Like Self-Charge Universal Module (SUM) combine energy harvesting and storage, shown in **Figure 5**. This compact system efficiently converts mechanical energy into electrical energy, capable of powering GPS modules for 30 minutes after 10 minutes of jogging [20].

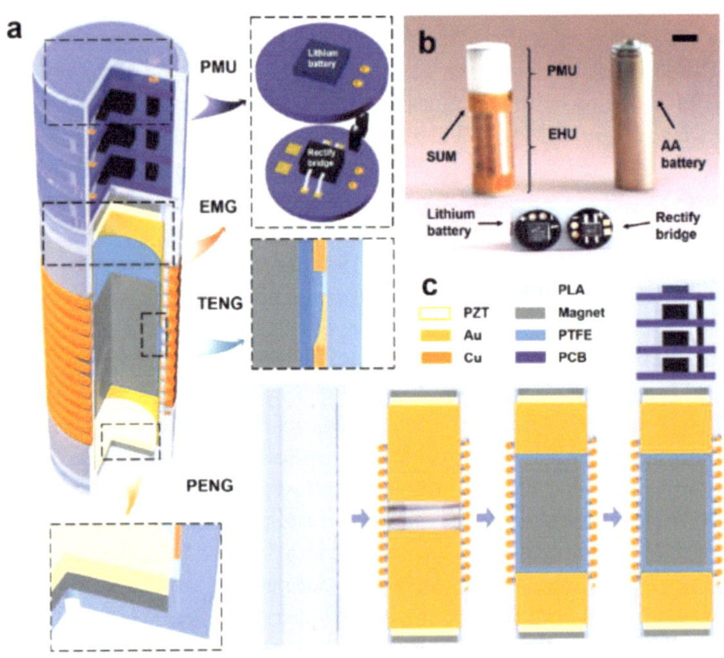

Figure 5.
(a) Structure of the SUM, (b) size of a SUM compared to a standard AA Battery, and (c) fabrication process of a SUM [20].

5. Challenges and limitations

Energy harvesting systems for wearable devices and satellite positioning technologies have demonstrated significant potential to revolutionize autonomy and sustainability. However, the practical implementation of these systems faces numerous challenges that must be addressed to ensure widespread adoption. One of the primary challenges in energy harvesting is the intermittent and unpredictable nature of ambient energy sources. For wearable devices, kinetic energy harvesters often struggle with irregular motion patterns. For example, many kinetic energy harvesting systems underperform at the low frequencies typically associated with human motion, necessitating optimization to improve real-world usability [21]. Similarly, solar energy systems for wearables face reduced efficiency in low-light or indoor conditions, making them unreliable as standalone solutions [22]. Energy harvesting technologies often provide lower energy densities compared to traditional batteries, limiting their ability to power high-demand systems. For instance, advanced hybrid harvesters can struggle to meet the energy requirements of devices with prolonged operation or high-power consumption, necessitating careful power budgeting and the use of small, low-power electronics [23]. Achieving high conversion efficiency across diverse energy sources is another technical barrier. Wearable systems must balance conversion efficiency with user comfort, size constraints, and operational reliability. For instance, piezoelectric energy harvesters for body-worn devices often exhibit narrow operating bandwidths, limiting their ability to harvest energy from random human motions effectively [24]. The miniaturization and integration of energy harvesters with existing systems remain significant hurdles. Flexible materials and compact designs are crucial for wearable applications but introduce complexities in manufacturing and cost. Moreover, integrating multiple energy harvesting mechanisms, such as combining piezoelectric and thermoelectric technologies, requires sophisticated energy management systems to handle varying energy outputs efficiently [25]. Finally, wearable energy harvesting systems [26] must use lightweight, flexible, and durable materials to ensure user comfort. However, these requirements increase production costs and complicate large-scale deployment. Additionally, the environmental impact of sourcing advanced materials for harvesters, such as rare-earth elements in piezoelectric systems, raises concerns about long-term sustainability [22].

6. Conclusion

Advancements in energy harvesting technologies have brought transformative potential to wearable devices and satellite positioning systems, contributing to a more sustainable and efficient future. This work explored the innovative applications of energy harvesting across various domains, focusing on combat boots, mobile satellite positioning systems, emerging materials, and the challenges these technologies face. The energy harvesting combat boot exemplifies how wearable devices can generate usable power from human motion. Utilizing air bulb-turbine systems, these boots efficiently convert kinetic energy into electricity, offering significant value for military, outdoor, and remote applications. By reducing reliance on traditional power sources, such technologies empower the operation of essential devices like GPS modules without external batteries, enhancing both functionality and user autonomy in the field. Expanding on this, other energy harvesting methods, such as piezoelectric materials, triboelectric nanogenerators, and advanced systems like carbon nanotube yarn harvesters, offer compact and efficient solutions for powering small devices. Emerging materials, including electroactive polymers and 2D materials, promise to further improve the performance, adaptability, and scalability of wearable energy technologies. These advancements underscore the versatility of energy harvesting systems in catering to diverse application scenarios, from personal electronics to industrial monitoring. The integration of energy harvesting with satellite positioning systems highlights another critical area of progress. Innovations such as insole-type triboelectric nanogenerators paired with Bluetooth beacons showcase how sustainable power solutions can meet the needs of modern navigation and tracking applications. Hybrid systems that combine GPS with Wi-Fi and microwave energy harvesting exemplify the adaptability of these technologies in reducing energy consumption while maintaining precision. Thermoelectric generators and optimized thermoelectric systems for satellites further illustrate the broader applicability of energy harvesting in aerospace and remote sensing contexts. Despite these advancements, significant challenges remain. Energy harvesting systems must address issues like intermittent energy sources, limited power density, material costs, and integration complexities to achieve greater reliability and usability. Overcoming these limitations will require continued research into hybrid systems, advanced materials, and energy management solutions. In conclusion, energy harvesting technologies bridge the gap between operational efficiency and environmental sustainability. From wearable combat boots to self-powered satellite positioning systems, these innovations are shaping a more self-reliant and eco-friendly future. By advancing the intersection of energy harvesting and modern technology, this field will continue to pave the way for smarter, more resilient systems capable of meeting the growing energy demands of diverse applications.

Author details

Tee Hui Teo[1*], Chiang Liang Kok[2], Xianlong Zhang[2] and Jovan Heng[2]

1 Singapore University of Technology and Design, Singapore

2 The University of Newcastle, Australia

*Address all correspondence to: tthui@sutd.edu.sg

References

[1] Akay H, Xu R, Han DCX, Teo TH, Kim S-G. Energy harvesting combat boot for satellite positioning. Micromachines. 2018;**9**(5):17. DOI: 10.3390/mi9050244

[2] Khaligh A, Zeng P, Zheng C. Kinetic energy harvesting using piezoelectric and electromagnetic technologies—state of the art. IEEE Transactions on Industrial Electronics. 2010;**57**(3):850-860. DOI: 10.1109/TIE.2009.2024652

[3] Othman A. Modeling of piezoelectric energy harvesting system embedded in soldier's boot using Matlab/Simulink. In: International Conference on Military Technologies (ICMT), Brno, Czech Republic. 2017. pp. 787-792. DOI: 10.1109/MILTECHS.2017.7988862. Available from: https://ieeexplore.ieee.org/document/7988862

[4] Kok CL, Fu X, Koh YY, Teo TH. A novel portable solar powered wireless charging device. Electronics. 2024;**13**:403. DOI: 10.3390/electronics13020403

[5] Choi Y-M, Lee MG, Jeon Y. Wearable biomechanical energy harvesting technologies. Energies. 2017;**10**(10):1483. DOI: 10.3390/en10101483

[6] Kim SH, Haines CS, Li N, Kim KJ, Mun TJ, Choi C, et al. Harvesting electrical energy from carbon nanotube yarn twist. Science. 2017;**357**(6353):773-778. DOI: 10.1126/science.aam8771

[7] Kornbluh RD, Eckerle J, McCoy B. A scalable solution to harvest kinetic energy. SPIE Newsroom. 2011. DOI: 10.1117/2.1201106.003749. Available from: https://spie.org/news/3749-a-scalable-solution-to-harvest-kinetic-energy

[8] Mokhtari F, Shamshirsaz M, Latifi M, Foroughi J. Nanofibers-based piezoelectric energy harvester for self-powered wearable technologies. Polymers. 2020;**12**(11). DOI: 10.3390/polym12112697

[9] Voss TJ, Subbian V, Beyette FR. Feasibility of energy harvesting techniques for wearable medical devices. In: 36th Annual International Conference of the IEEE Engineering in Medicine and Biology Society, Chicago. 2014. DOI: 10.1109/EMBC.2014.6943669

[10] Lee MH, Wu W. 2D materials for wearable energy harvesting. Advanced Materials Technologies. 2022;**7**(9). DOI: 10.1002/admt.202101623

[11] Xu C, Song Y, Han M, Zhang H. Portable and wearable self-powered systems based on emerging energy harvesting technology. Microsystems & Nanoengineering. 2021;**7**(25):17. DOI: 10.1038/s41378-021-00248-z

[12] Muller H, Randell C, Moss A. A 10mW wearable positioning system. In: 10th IEEE International Symposium on Wearable Computers, Montreux. 2006. DOI: 10.1109/ISWC.2006.286341

[13] Iadarola G, Disha D, Santis AD, Spinsante S, Gambi E. Global positioning system measurements: Comparison of IoT wearable devices. In: 2022 IEEE 9th International Workshop on Metrology for AeroSpace (MetroAeroSpace), Pisa. 2022. DOI: 10.1109/MetroAeroSpace54187.2022.9855994

[14] Martinsson W, Bagge R. Energy Efficient Geo-Localization for a Wearable Device. Lund University Publications [thesis]; 2015. Available from: https://lup.lub.lu.se/

student-papers/record/7451352; https://lup.lub.lu.se/student-papers/record/7451352/file/7451353.pdf

[15] Wang Z, Liu G, Cao J, Fu X, Fan B, Qin Y, et al. Self-powered position monitoring system based on insole-type wearable triboelectric nanogenerator and bluetooth beacon. Advanced Materials Technologies. 2023;**8**(18). DOI: 10.1002/admt.202300487

[16] Takacs A, Aubert H, Despoisse L, Fredon S. Microwave energy harvesting for satellite applications. Electronics Letters. 2013;**49**(11):722-724. DOI: 10.1049/el.2013.0372

[17] Lukowicz MV, Abbe E, Schmiel T, Tajmar M. Thermoelectric generators on satellites—an approach for waste heat recovery in space. Energies. 2016;**9**(7). DOI: 10.3390/en9070541

[18] Bierschenk J. Optimized thermoelectrics for energy harvesting applications. In: 17th IEEE International Symposium on the Applications of Ferroelectrics, Sante Re. 2008. DOI: 10.1109/ISAF.2008.4693950

[19] Bakır M, Karaaslan M, Karadağ F, Ünal E, Akgöl O, Alkurt FÖ, et al. Metamaterial-based energy harvesting for GSM and satellite communication frequency bands. Optical Engineering. 2018;**57**(8). DOI: 10.1117/1.OE.57.8.087110

[20] Tan P, Zheng Q, Zou Y, Shi B, Jiang D, Qu X, et al. A battery-like self-charge universal module for motional energy harvest. Advanced Energy Materials. 2019;**9**(36). DOI: 10.1002/aenm.201901875

[21] Magno M, Kneubühler D, Mayer P, Benini L. Micro kinetic energy harvesting for autonomous wearable devices. In: International Symposium on Power Electronics, Electrical Drives, Automation and Motion (SPEEDAM), Amalfi. 2018. DOI: 10.1109/SPEEDAM.2018.8445342

[22] Hesham R, Soltan A, Madian A. Energy harvesting schemes for wearable devices. AEU - International Journal of Electronics and Communications. 2021;**138**:2021. DOI: 10.1016/j.aeue.2021.153888

[23] Gunduz D, Stamatiou K, Michelusi N, Zorzi M. Designing intelligent energy harvesting communication systems. IEEE Communications Magazine. 2014;**52**(1):210-216. DOI: 10.1109/MCOM.2014.6710085

[24] Don Chua WF, Lim CL, Koh YY, Kok CL. A novel IoT photovoltaic-powered water irrigation control and monitoring system for sustainable city farming. Electronics. 2024;**13**:676. DOI: 10.3390/electronics13040676

[25] Zhu D, Wang J, Siek L, Kok CL, Qiu L, Zheng Y. High accuracy time-mode duty-cycle-modulation-based temperature sensor for energy efficient system applications. In: 2014 International Symposium on Integrated Circuits (ISIC), Singapore. 2014. pp. 400-403. DOI: 10.1109/ISICIR.2014.7029502

[26] Kok C-L, Siek L, Gao F, Zheng Y, Lim WM. An ultra-compact green bio-regulator dedicated for brain cortical implant using a dynamic PSR enhancement technique. In: 2012 Annual International Conference of the IEEE Engineering in Medicine and Biology Society, San Diego, CA, USA. 2012. pp. 1647-1650. DOI: 10.1109/EMBC.2012.6346262

Chapter 2

Perspective Chapter: Energy Harvesting in Wireless Communication – Radio Frequency and Visible Light Communication

Moslem Forouzesh and Mohsen Hossein Zadeh

Abstract

This chapter explores the integration of energy-harvesting techniques in wireless communication systems, focusing on two prominent methods: radio frequency (RF) energy harvesting and visible light communication (VLC) energy harvesting. RF energy harvesting leverages ambient electromagnetic waves from sources such as cellular networks, Wi-Fi, and television broadcasts to convert RF signals into usable electrical energy, providing a sustainable power solution for low-energy devices in applications like wireless sensor networks and the Internet of Things (IoT). The chapter discusses the architecture of RF energy-harvesting systems, including antennas, rectifiers, and energy storage units, highlighting their efficiency and adaptability in various environments. In contrast, VLC energy harvesting utilizes light emitted from LEDs and other light sources to generate power through photovoltaic cells. This method not only facilitates energy collection but also enables simultaneous data transmission, thereby enhancing the overall functionality of communication systems. The chapter examines the advantages and challenges associated with VLC, such as light intensity variations and low conversion efficiency. By comparing these two energies harvesting techniques, we aim to provide insights into their complementary roles in developing self-sustaining wireless communication networks that can operate efficiently in diverse conditions while minimizing reliance on traditional power sources.

Keywords: energy harvesting, radio frequency, visible light communication, wireless communication, wireless network

1. Introduction

Energy harvesting (EH) is a revolutionary technology that enables wireless devices to sustain their operation by harvesting energy from ambient sources like radio waves and light. This is particularly crucial for Internet of Things (IoT) devices and sensor networks, where battery replacement can be impractical or impossible. Energy-harvesting devices offer several significant advantages over conventional grid-powered and non-rechargeable battery-powered devices [1]. These advantages include energy self-sufficient, energy self-sustaining operation with lifetimes limited only by the lifetimes of their hardware. These devices utilize energy harvested from alternative natural resources such as solar, vibrational, electromagnetic, and thermoelectric. As such, their widespread adoption will reduce the use of conventional energy and the accompanying carbon footprint, and benefit the environment. In addition, these devices do not require conventional recharging, enabling untethered mobility, and deployment of these devices in hard-to-reach places such as remote rural areas, within structures, and within the human body.

In the last few decades, wireless communications have been in a phenomenal progress. They have brought extraordinary convenience in our everyday lives. A wireless data transmission does not need wires or cables. However, conventional wireless systems are limited by the batteries used to power wireless nodes. These have an impact on their mobility or their lifetime and thus prevent them from being deployed in increasingly widespread applications.

Energy harvesting from electromagnetic (EM) waves in wireless communication involves capturing and converting energy from ambient electromagnetic signals, such as radio waves, microwaves, and light, into electrical energy to power wireless devices and systems. This technique is especially useful in creating self-sustaining systems where replacing or recharging batteries frequently is difficult or impractical. In the following, we explain the main application of energy harvesting from electromagnetic waves in the wireless communication such as wireless sensor networks (WSNs), Internet of Things (IoT), body area networks.

In WSNs, sensors deployed in remote locations or harsh environments, such as agriculture, environmental monitoring, or industrial settings, can harvest energy from ambient EM waves. Sources include cellular signals, wireless fidelity (Wi-Fi), or TV broadcasts. These sensors can operate for extended periods without battery replacements, making them ideal for long-term monitoring applications. Moreover, in smart city applications, EM energy-harvesting sensors can monitor critical infrastructure, such as bridges, roads, or buildings, by harvesting energy from local wireless networks or radio towers [2].

In homes equipped with Wi-Fi or other wireless communication systems, IoT devices such as smart thermostats, lights, or security systems can harvest energy from electromagnetic signals to power themselves, reducing the need for constant battery changes or recharging. Furthermore, in industrial environments, EM energy-harvesting IoT sensors can monitor machinery, environmental conditions, or production processes by harvesting RF signals from wireless communication infrastructure or industrial radio systems.

Simultaneous wireless information and power transfer (SWIPT) systems allow wireless devices to receive both power and data through the same electromagnetic signals. This is useful for low-power devices such as RFID tags or wireless sensors that can harvest energy from communication signals while simultaneously transmitting data. Also, in future cellular networks, energy harvesting from high-frequency EM waves (such as those used in 5G and beyond) can enable IoT devices to power themselves while maintaining connectivity.

Radio frequency identification (RFID) tags use energy harvested from the electromagnetic waves emitted by RFID readers to transmit data. This makes them ideal for applications such as inventory tracking, access control, and supply chain management, where tags must operate without batteries. Moreover, near field communication (NFC) devices, commonly used for contactless payments and data exchange, can harvest energy from the electromagnetic field generated by NFC readers to power short-range wireless communication.

In the body area networks (BANs), wearable health monitoring devices, such as fitness trackers, heart rate monitors, and glucose sensors, can harvest energy from electromagnetic waves, including signals from mobile phones, Bluetooth devices, or Wi-Fi routers. This helps reduce the need for recharging, making these devices more convenient for users. It is necessary to mention that energy harvesting from EM waves can power implantable medical devices, such as pacemakers or neural stimulators, by capturing energy from external wireless communication systems. This can extend the life of implants and reduce the need for surgical battery replacements.

Sensors deployed in smart cities to monitor traffic flow, street lighting, or environmental conditions can harvest energy from electromagnetic signals emitted by communication towers, public Wi-Fi networks, or nearby mobile devices. Moreover, surveillance cameras, air quality sensors, and public safety devices in smart cities can be powered through energy harvested from EM waves, making them more sustainable and reducing maintenance costs.

Wearable devices such as smartwatches, fitness trackers, and health monitoring devices can harvest energy from nearby EM sources, such as mobile phones, Wi-Fi routers, or Bluetooth devices, enabling continuous operation without frequent charging. Smart fabrics or clothing with embedded sensors can harvest energy from ambient electromagnetic waves, enabling continuous health monitoring or providing communication functionalities.

Sensors used in agriculture, environmental monitoring, or industrial automation can be powered by harvesting energy from electromagnetic waves, such as those emitted by radio towers or communication systems, allowing for continuous operation in remote or hazardous locations. Furthermore, low-power communication nodes in wireless networks can harvest energy from nearby wireless signals (e.g., Wi-Fi, 4G/5G signals) to sustain themselves, creating energy-efficient and self-sustaining network architectures.

Unmanned aerial vehicles (UAVs) and drones used for surveillance, package delivery, or environmental monitoring can harvest energy from electromagnetic waves, extending their flight time and reducing the need for frequent battery replacements or recharging [3]. In urban areas, they can capture energy from the dense wireless communication infrastructure. Drones used in agriculture, construction, or disaster recovery can rely on electromagnetic wave energy harvesting to power sensors and communication systems, allowing for longer and more efficient operation.

Energy harvesting from electromagnetic waves including radio frequency and visible light communication can be applied to underwater or marine environments to power sensors that monitor water quality, marine life, or oceanographic data. These sensors can harvest energy from surface communication signals or nearby infrastructure. Note that underwater communication systems can utilize energy harvested from surface signals to power autonomous underwater vehicles or sensors used for deep-sea exploration.

Cellular network base stations and relay nodes can use electromagnetic wave energy harvesting to reduce their energy consumption. By capturing ambient energy from nearby communication systems or the environment, base stations can contribute to more sustainable and energy-efficient communication networks.

The energy harvesting is important technique in military and defense Applications. Wearable communication devices, health monitors, or sensors used by

Applications	Icon	Energy source
WSNs	Sensor node	Solar, RF signals
IoT	Connected home with smart devices	RF signals from Wi-Fi, solar energy
RFID tag	RFID tag	RF signals from a reader
Wearable devices	Smartwatch	Body heat, movement
SWIPT	Wireless communication tower with energy and data symbols.	RF signals from cellular networks
Smart grids	Power meter	Solar power, RF signals
Drones and robotics	Drone	Solar, RF energy
Environmental monitoring	Environmental sensor (e.g., air quality monitor)	Solar, RF signals
Military applications	Soldier with wearables	RF signals, body heat

Table 1.
Applications of energy harvesting in wireless communication.

soldiers in the field can harvest energy from electromagnetic signals emitted by communication systems, reducing the need to carry extra batteries or power sources [4]. Military drones, unmanned ground vehicles (UGVs), and autonomous systems used for reconnaissance or surveillance can harvest energy from electromagnetic waves to extend operational time and improve mission capabilities.

Finally, we can say that energy harvesting from electromagnetic waves in wireless communication enables the development of self-powered, sustainable devices and systems across a wide range of applications. From smart cities and IoT devices to medical implants and military systems, harvesting energy from ambient RF, microwave, or visible light signals offers significant advantages in terms of reduced maintenance, increased device lifetime, and lower environmental impact. This technology is crucial for enabling next-generation wireless communication networks, supporting the growth of IoT, and promoting greener and more sustainable infrastructure. We categorize these applications in **Table 1**.

2. Limitations of energy harvesting in wireless communication

While energy harvesting promises greener, self-powered networks, it faces several limitations, which explain them in the following.

2.1 Low energy density

Energy sources in the environment, like RF signals, vibrations, or thermal gradients, typically provide only tiny amounts of energy. For example, RF signals can deliver power in the microwatt to milliwatt range, while devices used in wireless communication, such as sensors or smartphones, often need more significant power (ranging from milliwatts to watts) to function effectively. Since the amount of energy harvested is small, it is challenging to power devices continuously or perform energy-intensive tasks. This limits the range and types of applications that can rely on energy harvesting, often restricting it to low-power devices like basic sensors or small IoT devices that transmit small data packets intermittently.

2.2 Intermittent energy supply

Most energy-harvesting methods rely on ambient energy sources that are not always available or consistent. For instance, solar energy depends on the presence of sunlight, which can be blocked by weather conditions or day-night cycles. Similarly, RF energy requires nearby signal sources, which can vary in availability or strength based on distance or the surrounding environment. Since the availability of energy is unpredictable, wireless communication devices may not have a consistent or reliable power source. This intermittency can cause devices to shut down or operate erratically, leading to interrupted communication or data loss.

2.3 Low conversion efficiency

Conversion efficiency refers to how effectively the ambient energy can be turned into usable electrical power. For example, RF energy-harvesting systems use a component called a retenna (rectifying antenna), which captures RF signals and converts them into DC power. However, this process often suffers from efficiency losses, meaning only a fraction of the available energy is captured and converted. Hence, if ambient energy is available, a significant portion of it can be lost during the conversion process. For RF energy harvesting, the efficiency of converting RF signals into power typically ranges from 10 to 50%. This limits the total amount of usable power, further constraining the potential applications of the technology.

2.4 Mismatch between power supply and demand

The energy demand for different applications varies greatly. Wireless communication systems often need bursts of energy for transmitting data, processing, or maintaining network connectivity. On the other hand, energy harvesting generally provides a steady, but low, trickle of power. This mismatch between power supply (low, continuous) and demand (high, intermittent) creates operational challenges. Therefore, for high-power applications like streaming video or handling large amounts of data, energy harvesting alone cannot provide the necessary power. Even for lower-power applications like sensor networks, when these devices need to transmit data over long distances, their power needs may spike beyond what energy harvesting can supply. This limits the use of energy harvesting to devices with very low power requirements or forces designers to combine energy harvesting with batteries.

2.5 Energy storage requirements

Because ambient energy is often intermittent and insufficient, energy-harvesting devices frequently rely on storage mechanisms like batteries or capacitors to store excess energy for later use. However, energy storage systems come with their own limitations, such as capacity, efficiency, weight, and cost. Storing energy adds complexity to the system, increases costs, and can lead to faster wear and tear of storage components. Batteries degrade over time, reducing the overall lifespan of the system. Super capacitors, while faster at charging, tend to store less energy than batteries. Additionally, the inclusion of storage components increases the device's size, making it less suitable for small or portable applications.

2.6 Complexity in system design

Designing an energy-harvesting system requires carefully balancing multiple components—harvesting mechanisms (e.g., antennas or piezoelectric elements), conversion circuits, storage, and power management units. Each component must be designed to work efficiently without drawing too much power itself. This complexity increases the cost, size, and weight of the device. Moreover, it introduces trade-offs in terms of power efficiency, data transmission performance, and overall device functionality. For example, optimizing the antenna for RF energy harvesting may reduce its ability to handle communication tasks, creating a design challenge.

2.7 Regulatory and health concerns

Wireless energy transfer, particularly through RF signals, has raised concerns about the potential health effects of prolonged exposure to electromagnetic radiation. Regulatory bodies, such as the FCC, impose limits on how much RF power can be transmitted in public spaces to ensure safety. Therefore, to comply with safety regulations, the power transmitted for energy harvesting must be kept at safe levels, which limits the amount of energy that can be harvested by devices. Higher power levels would allow for more efficient energy harvesting but would violate safety standards, potentially affecting public health and increasing the risk of interference with existing communication systems.

Energy harvesting in wireless communication is still in its early stages, offering great potential for creating self-sustaining, battery-free systems. However, its current limitations—such as low-energy density, inconsistent supply, poor conversion efficiency, and the need for storage—mean that it is mostly suitable for low-power applications like simple sensors and low-data-rate IoT devices. Significant technological advancements are required before energy harvesting can reliably support more power-intensive applications, such as smartphones, large-scale sensor networks, or communication systems with high data throughput and long-range requirements.

3. Leading companies in the field of energy harvesting

3.1 Powercast corporation

Powercast specializes in wireless power transfer and energy-harvesting technologies. They provide solutions for powering wireless sensors and RFID tags using RF energy harvesting. Powercast's Powerharvester receivers capture RF energy from the environment, converting it into DC power that can be used to power IoT devices or recharge batteries wirelessly over distance.

3.2 Cypress semiconductor (now part of Infineon)

Cypress, now part of Infineon, has been involved in developing energy-harvesting solutions for low-power wireless communication in IoT applications. They focus on energy-efficient Bluetooth and wireless connectivity technologies. This company offers products like energy-harvesting power management ICs, which convert energy from sources such as solar cells or thermoelectric generators into usable power for wireless devices.

3.3 Texas instruments

Texas Instruments (TI) produces power management and energy-harvesting components that can be used in wireless telecommunications, especially for IoT applications. They work on ultra-low-power designs and energy-harvesting interfaces. TI's energy-harvesting power management ICs and wireless connectivity solutions (e.g., sub-GHz radios) enable self-sustaining IoT devices, reducing reliance on battery replacements.

3.4 EnOcean GmbH

EnOcean is a leader in self-powered IoT devices and energy-harvesting technology, particularly in the building automation, smart home, and industrial sectors. The company's wireless solutions utilize energy harvested from motion, light, or temperature changes. EnOcean energy-harvesting modules power wireless sensors and switches, eliminating the need for batteries and enabling maintenance-free operation.

3.5 Wiliot

Wiliot works on IoT solutions using ambient energy harvesting from radio frequencies (RF). They create battery-free sensors that operate using RF energy collected from the surrounding environment. Wiliot's IoT tags, which are powered by ambient RF energy, can transmit data over Bluetooth, enabling wireless communications without the need for traditional batteries.

These companies are recognized for their advanced technologies in energy harvesting and wireless communications, and their solutions are key drivers in the development of sustainable, long-lasting IoT networks.

4. Understanding visible light communication and its role in energy harvesting

Visible light communication (VLC) is a wireless communication technology that uses visible light (wavelengths between 400 and 800 mm) for transmitting data as depicted in **Figure 1**. It leverages existing light sources, such as light-emitting diode (LED) bulbs, to transmit information while illuminating the environment. VLC is often considered a part of light fidelity (Li-Fi), an alternative to Wi-Fi, where light is used for high-speed data transmission. In this technology, LEDs can switch on and off at extremely high speeds (imperceptible to the human eye) to send binary data. By modulating the intensity of light, the LED transmits information to a photodiode, which acts as the receiver. The receiver that is usually a photodiode or image sensor detects changes in light intensity and converts them back into electrical signals that are processed to retrieve the transmitted data. VLC can provide high data rates, up to several gigabit per second, and is particularly useful for indoor communication systems like smart homes, offices, and secure data transfer environments.

VLC can be used for precise indoor positioning, particularly in environments where GPS signals are weak or unavailable, such as airports, shopping malls, and museums. It provides location-based services with high accuracy. VLC enables smart lighting systems where LED lights are used for both illumination and data transmission. It plays a key role in smart homes, cities, and building automation, allowing

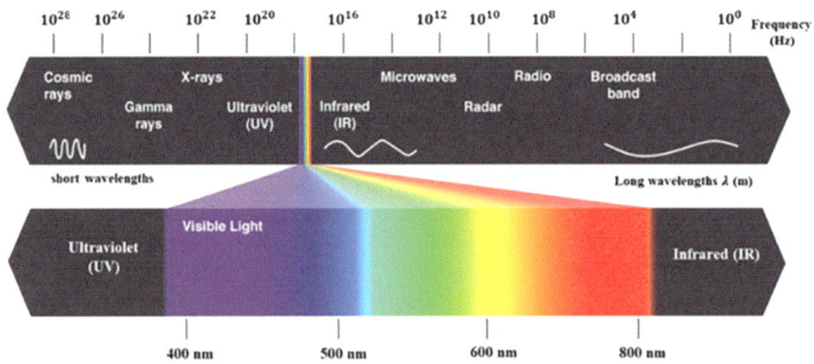

Figure 1.
Visible light wavelengths [5].

communication between IoT devices using existing lighting infrastructure. VLC can facilitate communication between vehicles (V2V) and between vehicles and traffic infrastructure (V2I). This technology enhances autonomous driving, smart traffic management, and road safety by using vehicle headlights and traffic lights for communication. VLC is used for Li-Fi, which enables high-speed, wireless internet through LED lighting. It offers higher data transmission rates than Wi-Fi and is ideal for secure, high-density environments such as homes, offices, and hospitals. VLC can facilitate wireless communication in hospitals, enabling data transfer between medical devices and supporting patient monitoring systems. Additionally, it allows real-time data transmission from wearable health devices without interfering with other essential equipment.

Energy harvesting in VLC involves capturing energy from visible light (typically emitted by LED lights used for communication) and converting it into usable electrical power. This enables devices, such as sensors or small electronics, to power themselves using the same light used for data transmission.

4.1 Light energy-harvesting components

As depicted in **Figure 2**, light energy-harvesting components refer to the essential devices used to capture and convert visible light into electrical energy for use in energy-harvesting systems such as photovoltaic cells and photodiodes. In the context of VLC, these components play a dual role by both harvesting energy and facilitating communication. To clear this item, we explain them in the following.

Photovoltaic cells convert light into electrical energy through the photovoltaic effect. When light hits the cell, it excites electrons, creating an electric current. In VLC systems, photovoltaic cells can harvest energy from visible light while not necessarily participating in data reception, especially in outdoor or high-light environments. Typically made of silicon or other semiconducting materials, these cells are optimized for harvesting light energy efficiently. Photodiodes are semiconductor devices that convert light into electrical current. They are commonly used in VLC systems as receivers for data transmission, but they can also be adapted to harvest energy. In energy-harvesting VLC systems, photodiodes can perform both tasks—receiving data and converting excess light into usable power. This dual functionality

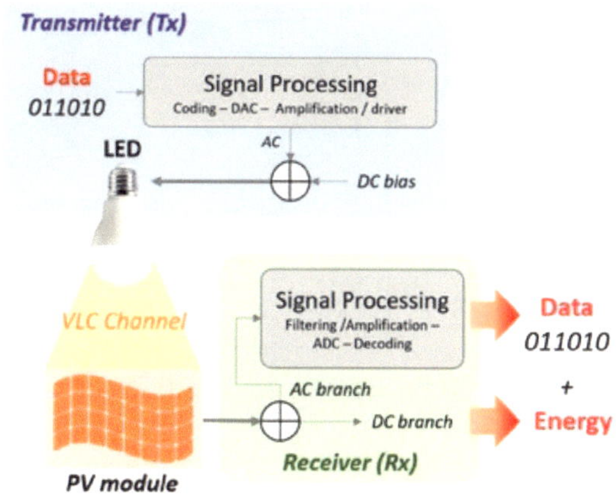

Figure 2.
Schematic representation of a VLC chain illustration [6].

reduces the need for separate components for communication and energy harvesting. Photodiodes are highly sensitive to changes in light intensity that is crucial for detecting data in VLC while simultaneously harvesting energy from ambient light [7].

In summary, energy harvesting in VLC involves collecting light energy from the same LED sources that transmit data, allowing for the creation of self-powered or low-power IoT devices. This concept integrates data transmission and power generation into a single system, which is particularly useful in IoT, smart buildings, and low-power communication networks.

4.2 Challenge of energy harvesting in VLC

4.2.1 Energy-harvesting efficiency

The amount of energy that can be harvested from visible light, especially indoor lighting, is relatively low. This limits the power available for operating devices, particularly for energy-hungry applications. Moreover, photovoltaic cells and photodiodes used for energy harvesting have limited efficiency in converting light into electrical energy, especially in low-light environments. Current technology is often not efficient enough to power more than small, low-power IoT devices or sensors. Moreover, in VLC, there is a potential conflict between energy harvesting and communication. If the same light source is used for both data transmission and energy harvesting, optimizing the system for one function (data transmission or energy harvesting) may degrade the other.

4.2.2 Line-of-sight requirement

As seen in **Figure 3**, VLC requires a clear line-of-sight(LOS) link between the transmitter (light source) and the receiver (photodiode or photovoltaic cell). Any obstacle blocking the light will disrupt both data transmission and energy harvesting. Shadows, furniture, or moving objects can interrupt the line-of-sight, leading to inconsistent communication and energy collection. This makes VLC and energy harvesting less reliable in dynamic or cluttered environments, such as offices or homes.

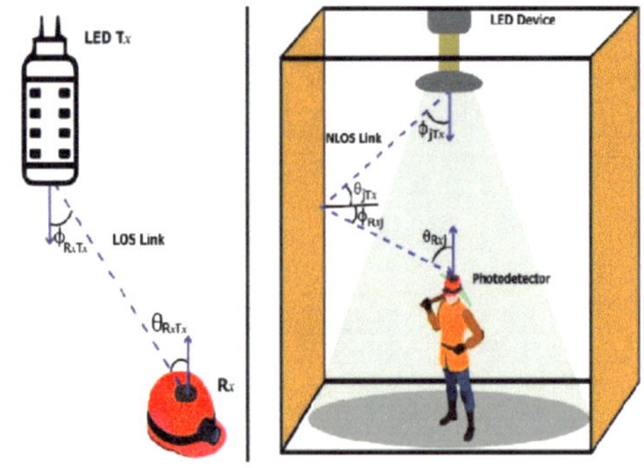

Figure 3.
Schematic representation of a VLC chain illustration [8].

4.2.3 Dependence on light conditions

The effectiveness of both VLC and energy harvesting is heavily dependent on the light intensity and ambient conditions. For instance, in dimly lit environments or at night, the performance of both systems declines significantly. Indoor lighting typically provides less energy compared to outdoor sunlight, limiting the potential for energy harvesting. Outdoor VLC systems face challenges with fluctuations in sunlight and environmental factors like weather conditions, which can affect performance.

It is necessary to mention these challenges must be addressed through advances in technology, efficient light-to-energy conversion, and smart system designs to make VLC and energy harvesting a viable alternative in communication and power systems.

5. Architecture of RF energy-harvesting network

RF energy-harvesting networks utilize ambient radio frequency signals to power low-energy devices, making them a vital component in the development of sustainable wireless communication systems. The architecture of these networks can be complex, involving multiple components that work together to efficiently capture and convert RF energy into usable electrical power. Below is a detailed exploration of the architecture and its components.

5.1 General architecture

The architecture of an RF energy-harvesting network typically includes:

5.1.1 Energy-harvesting zone

This area is designated for capturing RF energy from external sources, such as transmitters or broadcast stations.

5.1.2 Information transmission zone

This zone is where data communication occurs, allowing nodes to send and receive information.

The energy-harvesting zone is usually smaller than the information transmission zone because the operational power required for energy harvesting is significantly higher than that needed for information decoding [9].

5.2 Node structure

Each node in RF energy-harvesting networks consists of several key components as depicted in **Figure 4**.

5.2.1 RF energy harvester

- *Antenna*: Captures RF signals, designed for either single or multiple frequency bands. Multi-band antennas are advantageous as they can harvest energy from various sources simultaneously.

- *Impedance matching circuit:* Ensures optimal power transfer from the antenna to the rectifier by matching the impedance of the antenna with that of the rectifier.

- *Rectifier*: Converts the captured RF signals into direct current (DC). Advanced designs may include voltage doublets or full-wave rectifiers to enhance efficiency, achieving conversion efficiencies up to 71.5% [10].

- *Energy storage*: Typically consists of capacitors or rechargeable batteries that store harvested energy for later use.

5.2.2 Microcontroller

A low-power microcontroller manages data processing and controls the operation of other components based on energy availability.

5.2.3 Power management module

This module decides how harvested energy is utilized:

- *Harvest-use mode*: The harvested energy is used immediately to power the node.

- *Harvest-store-use mode*: Excess energy is stored in a battery for future use, allowing continuous operation even when harvesting conditions are poor [9].

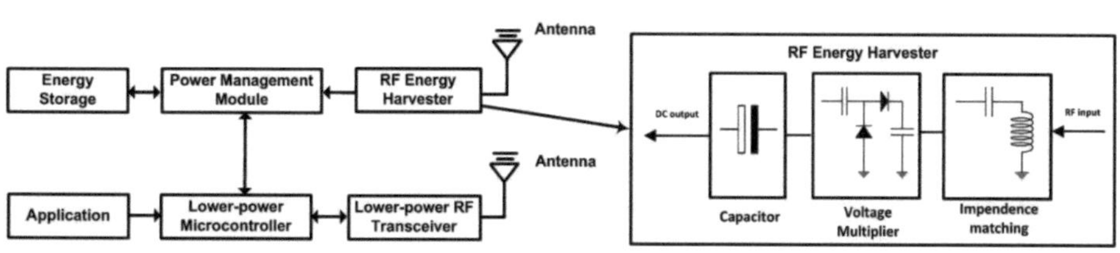

Figure 4.
A general architecture of an RF energy-harvesting device [9].

5.3 Communication protocols

Efficient communication protocols are essential for managing energy demands and ensuring reliable data transmission. Some protocols include:

5.3.1 Request-for-energy (RFE) protocols

Low-energy sensors send RFE packets to request energy from peripheral transmitters, which respond with cleared-for-energy (CFE) packets, allowing for coordinated energy distribution based on demand [10].

5.4 Architectural variations

To enhance performance, various architectural designs have been developed:

5.4.1 Multi-port systems

These systems incorporate multiple antennas and rectifiers to capture RF energy from different sources, improving overall efficiency.

5.4.2 Dual-band systems

Utilizing dual-band antennas allows simultaneous harvesting from two frequency bands (e.g., ISM 2.45 GHz and ISM 5.8 GHz), which can significantly increase the amount of harvested energy [10].

5.5 Simultaneous wireless information and power transfer (SWIPT)

Simultaneous wireless information and power transfer (SWIPT) is a groundbreaking technology that enables the concurrent transmission of data and energy over the same wireless medium. This dual functionality is particularly beneficial for battery-operated devices, such as those found in the Internet of Things (IoT) and wireless sensor networks (WSNs), where both power and communication capabilities are essential. A typical SWIPT system consists of a transmitter that generates RF signals carrying both information and power, and a receiver that captures this energy while simultaneously extracting the transmitted data. Techniques such as Orthogonal Frequency Division Multiplexing (OFDM) and transmit beamforming enhance the efficiency of resource sharing, allowing for optimal performance in varying channel conditions [11].

Despite its advantages, implementing SWIPT presents several challenges, including trade-offs between data rate and power transfer efficiency, as well as hardware limitations in designing receivers capable of simultaneous energy harvesting and data demodulation. Nevertheless, SWIPT has promising applications across various fields, including wireless sensor networks, IoT devices, wearable technology, and device-to-device communication. By providing continuous power supply while enabling reliable data transmission, SWIPT paves the way for more sustainable and efficient wireless networks, significantly enhancing the operational longevity of low-power devices in diverse applications [11].

6. RF energy-harvesting technique

RF energy-harvesting systems can be classified into two main types:

6.1 Dedicated RF energy harvesters

Dedicated RF energy harvesters are specialized systems designed to capture and convert radio frequency (RF) signals from dedicated RF sources into usable electrical energy. These systems are particularly advantageous in applications requiring a predictable and controllable energy supply, as they utilize fixed transmitters that emit RF signals specifically for energy harvesting. Typically operating within license-free ISM frequency bands, dedicated RF sources can provide a stable energy output, making them ideal for powering devices in environments where ambient RF energy may be insufficient or unreliable. The architecture of these harvesters includes essential components such as an antenna for capturing RF signals, an impedance matching circuit to optimize power transfer, a rectifier to convert AC signals to DC, and an energy storage unit to hold the harvested energy for later use.

One of the main benefits of dedicated RF energy harvesters is their ability to ensure consistent power delivery, which is crucial for devices with strict quality of service (QoS) requirements. However, deploying such systems can be costly due to the need for infrastructure investment in dedicated RF transmitters. Moreover, regulatory limitations on output power can restrict the amount of energy transmitted, necessitating multiple transmitters to meet user demands effectively. Despite these challenges, dedicated RF harvesters offer a reliable solution for powering low-energy devices in various applications, including wireless sensor networks and IoT devices, by providing a sustainable and efficient energy source that minimizes reliance on traditional batteries.

6.2 Ambient RF energy harvesters

Ambient RF energy harvesters are systems designed to capture and convert radio frequency (RF) signals from the surrounding environment into usable electrical energy. Unlike dedicated RF energy harvesters that rely on specific transmitters, ambient RF harvesters utilize naturally occurring RF signals emitted by various sources, such as Wi-Fi routers, cellular networks, and television broadcasts. The typical architecture of an ambient RF energy-harvesting system includes a receiving antenna that captures the RF signals, an impedance matching circuit to optimize power transfer, a rectifier to convert the captured AC signals into DC power, and an energy storage unit to store the harvested energy for later use. The efficiency of these systems heavily depends on the design of each component, particularly the antenna's gain and the effectiveness of the impedance matching circuit, which minimizes losses during energy conversion. One of the significant advantages of ambient RF energy harvesting is its ability to operate in diverse environments without requiring dedicated infrastructure. This technology is particularly beneficial for powering low-energy devices in applications such as wireless sensor networks (WSNs) and Internet of Things (IoT) devices, where long-term battery life is crucial. However, challenges remain, including the relatively low power density available from ambient sources compared to dedicated sources, which can limit the performance of these systems. Additionally, variations in signal strength due to distance from RF sources and environmental factors can affect harvesting efficiency. Despite these challenges, ambient RF energy harvesting presents a promising solution for sustainable energy supply in modern wireless applications, allowing devices to operate autonomously without frequent battery replacements [9, 12].

7. Conclusion

In conclusion, the integration of energy-harvesting techniques, specifically radio frequency (RF) energy harvesting and visible light communication (VLC) energy harvesting, marks a significant advancement in sustainable power solutions for wireless communication systems. RF energy harvesting effectively captures ambient electromagnetic waves, providing a reliable energy source for low-power devices in applications such as the Internet of Things (IoT) and wireless sensor networks, while its adaptability to diverse environments minimizes the need for frequent battery replacements. Conversely, VLC energy harvesting not only generates power through light but also facilitates simultaneous data transmission, enhancing overall system functionality. By leveraging the strengths of both techniques, a hybrid approach could optimize efficiency and sustainability in future wireless communication networks, ultimately contributing to an eco-friendlier technological landscape that reduces reliance on traditional energy sources.

Author details

Moslem Forouzesh* and Mohsen Hossein Zadeh
Faculty of Engineering Modern Technologies, Amol University of Special Modern Technologies, Amol, Iran

*Address all correspondence to: m.forouzesh@ausmt.ac.ir

References

[1] Raghunathan V, Ganeriwal S, Srivastava M. Emerging techniques for long lived wireless sensor networks. IEEE Communications Magazine. 2006;**44**:108-114. DOI: 10.1109/MCOM.2006.1632657

[2] Grossi M. Energy harvesting strategies for wireless sensor networks and mobile devices: A review. Electronics. 2021;**10**:1-55. DOI: 10.3390/electronics10060661

[3] Nguyen MT, Nguyen CV, Truong LH, Le AM, Quyen TV, Masaracchia A, et al. Electromagnetic field based wpt technologies for uavs: A comprehensive survey. Electronics. 2020;**9**(3):461

[4] Sisto J. Soldiers of the future will generate their own power. 2014. Available from: www.army.mil, https://www.army.mil/article/138057/Soldiers_of_the_future_will_generate_their_own_power

[5] El Jbari M, Moussaoui M. High-performance metric of graphene-based heterojunction LEDs and PDs in visible light communication systems. In: Recent Advances in Graphene Nanophotonics. Springer Nature Switzerland: Cham; 2023. pp. 31-53

[6] Bouclé J, Ribeiro Dos Santos D, Julien-Vergonjanne A. Doing more with ambient light: Harvesting indoor energy and data using emerging solar cells. In: Solar. Vol. 3. MDPI; 2023. pp. 161-183

[7] Shore A, Roller J, Bergeson J, Hamadani BH. Indoor light energy harvesting for battery-powered sensors using small photovoltaic modules. Energy Science & Engineering. 2021;**9**:2036-2043. DOI: 10.1002/ese3.964

[8] Javaid F, Wang A, Sana MU, Husain A, Ashraf I. Characteristic study of visible light communication and influence of coal dust particles in underground coal mines. Electronics. 2021;**10**(8):883

[9] Lu X, Wang P, Niyato D, Kim DI, Han Z. Wireless networks with RF energy harvesting: A contemporary survey. IEEE Communications Surveys & Tutorials. 2014;**17**:757-789. DOI: 10.1109/COMST.2014.2368999

[10] Benkalfate C, Ouslimani A, Kasbari AE, Feham M. A new RF energy harvesting system based on two architectures to enhance the DC output voltage for WSN feeding. Sensors. 2022;**22**:3576. DOI: 10.3390/s22093576

[11] Liang Y, He Y, Qiao J, Hu AP. Simultaneous wireless information and power transfer in 5G mobile networks: A survey. In: Proceedings of the Computing, Communications and IoT Applications (ComComAp); 26-28 October 2019; Shenzhen, China: IEEE. 2019. pp. 460-465. DOI: 10.1109/ComComAp46287.2019.9018837

[12] Khan NU, Khan FU, Farina M, Merla A. RF energy harvesters for wireless sensors, state of the art, future prospects and challenges: A review. Physical and Engineering Sciences in Medicine. 2024;**47**:385-401

Chapter 3

Perspective Chapter: Joint Energy Harvesting and Data Transfer for Medical Implants Exploiting Light – Concept and Preliminary Results

Syifaul Fuada, Mariella Särestöniemi and Marcos Katz

Abstract

The potential of using light as an alternative energy source and a medium for communicating with medical implants is attractive. Light can propagate through biological tissues, and it is well known that maximum penetration takes place with near-infrared (NIR) light. Light has several advantages, making it quite attractive for communication compared to radio frequency (RF) or acoustic waves. Light communications, in particular, are highly secure, safe, and private wireless links while also presenting opportunities for low-power implementation. Moreover, the light-based operation is free from electromagnetic interference. In the context of the wireless power transfer (WPT) method for medical implants, light also has unique advantages compared to RF and acoustics waves, that is, secure wireless charging capabilities. The emission of NIR light carrying the modulated data that passes through the biological tissue can further be converted into electrical energy by photovoltaic (PV) cells, enabling medical implants to be powered by light. In this chapter, a new paradigm in modern implanted devices, that is, joint data transfer and energy harvesting by exploiting light, is described. We present the progress of research (preliminary results) on experiments using a tissue-mimicking phantom emulating the human soft tissue, conducted in our laboratory to prove the concept, showcasing the potential of the data transfer and energy harvesting method by utilizing light. We demonstrate that optical channels are established at a depth of approximately 40 mm across the phantom, allowing for both wireless data and energy transfer with the implanted devices.

Keywords: energy harvesting, wireless data transfer, near-infrared light, biological tissue, phantom, implanted device, implant communications, wireless energy transfer

1. Introduction

Electronic medical implants have been shown to have a positive impact on transforming healthcare [1]. Electrical medical implants are one of the in-body device types [2]; they are tiny devices placed shallower (e.g., subcutaneously) or deeper inside the human body at particular locations to perform various operations; they are typically used for sensing (e.g., monitoring, measuring) and actuation (delivering treatment). In both cases, they need to be connected to the outside world. Medical implants play a vital role in the wireless body area network (WBAN) concept, which is a healthcare paradigm connecting either implanted or wearable devices through sensor nodes to external command centers [3]. Examples of medical implants include pacemakers, defibrillators, cardiovascular pressure monitors, intra-cranial pressure monitors, deep brain neurosensors, retina stimulators, chronic pain stimulators, drug infusion systems, identity verification chips, glucose monitors, cochlear implants, and so on [4]. In a WBAN setting, a standard communication method is required to wirelessly connect medical implants to a device located outside the human body (out-body device) [5]. Radio frequency (RF) and acoustic waves are classical communication methods typically used to provide wireless connectivity in WBAN [6].

In another context, medical implants are energy-limited, as they are typically powered by batteries, which necessitate frequent recharging to prolong the battery life [7]. Most medical implants nowadays use non-rechargeable batteries. Non-rechargeable battery depletion frequently requires replacing the implant's battery through surgery procedure [8], which can be costly (around USD 6000–12,000) [9], carry a risk of complications [10], and cause stress for patients [11] and other negative impacts. Given the urgent need to prolong the medical implant's battery, a solution leveraging technology is needed.

With the growing reliance on data in healthcare, ensuring the safe exchange of patient information is more important than ever [12, 13]. Exploring light to communicate wirelessly with implanted devices is particularly promising as, due to its unique advantages, it can be used as an alternative to RF and acoustic communications for implanted devices [14]. Wireless communication approaches utilizing light are known as optical wireless communications (OWC). OWC offers several benefits, that is, high security for sensitive patient information, considered safe, and interference-free operation [15]. While OWC holds great promise for securing patient information because of the inherent nature of light, it is not without challenges. Light is typically scattered, absorbed, and reflected when it interacts with biological tissues [16]. Choosing an appropriate wavelength of light is necessary for OWC to work effectively when communicating with an implanted device. Near-infrared (NIR) light has shown to be particularly useful for this purpose as it is less affected by the scattering and absorption effect in the tissue that can weaken optical signals, making it pass through biological tissues more effectively than other light wavelengths [17]. For this reason, NIR light is considered a strong candidate for secure communication to implanted devices.

In the context of wireless powering implanted devices, energy harvesting (EH) techniques are a promising solution. Different approaches have been investigated, including harvesting energy from external sources, where RF and acoustics, besides being used as communication methods, are typically employed for this purpose [18]. However, the utilization of light, especially NIR light, for power transfers has emerged as a particularly intriguing prospect where safe energy charging and interference-free operation can be compromised when dealing with security aspects.

In this chapter, we demonstrate the potential of OWC for in-body communication, particularly focusing on NIR light 810 nm, as it can be considered effective in penetrating deeper into biological tissue. A photodetector was used to receive the data from the transmitted NIR light. This method is expected to create safer, more efficient ways to manage medical devices and protect patients' private health information. In order to harvest energy from the same light source, which is already used for data communication, a PV cell integrated with an energy harvesting unit is employed. Leveraging PV cells to capture and convert light energy offers the potential to extend the operational lifespans of medical implants while simultaneously enabling secure data transfer. The dual-functionality approach is projected to transform future healthcare into something better than ever.

The remainder of this chapter is structured as follows: Section 2 describes RF, acoustics, and light as wireless link enabler technologies for implanted devices, followed by a discussion on their advantages and disadvantages; Section 3 discusses their specific functions in WPT specifically for implanted devices, followed by their benefits and drawbacks. Section 4 elaborates on the concept of joint energy harvesting and data transfer for medical implants exploiting light and demonstrates the results of preliminary experimental studies conducted in our laboratory. Finally, we present the conclusion in Section 5.

2. Data transmission method to communicate with medical implants

As medical technology continues to advance, researchers are increasingly looking for new ways to enable secure communication for implanted devices. Modern medical implants employ a wireless transceiver system for communication, which becomes a central functionality of the device [19]; it provides a non-intrusive connection method. Medical implants equipped with a wireless transmission capability can act as nodes in a WBAN environment. Designing these medical implants should meet specific standards for effective data transmission. To this end, numerous technologies, including acoustic, RF, and OWC, can establish this wireless link, each offering distinct advantages and disadvantages. The following subsection briefly reviews these technologies:

2.1 Radio frequency (RF) communication

RF refers to electromagnetic waves within the frequency range of 3 kHz to 300 GHz [20]. RF communication has been extensively used in all major applications, including providing wireless connectivity to medical implants. Traditionally, RF communication has been used to send and receive data to and from the implanted device. One advantage of RF is that it enables low-power wireless connections, which is vital for medical implants that are energy consumption-constrained [14]. Additionally, RF communication enables both non-contact sensors [21–23]. In 1999, the Medical Implant Communication Service (MICS) was introduced, using the 402–405 MHz RF spectrum. MICS allows wireless communication with medical implants over distances of up to 2 meters [24]. This RF band is essential for WBANs, which enable data transfer between external devices and medical implants. The specific spectrum allocation for MICS helps ensure that communication with medical implants is free from interference, even with surrounding RF-dependent medical devices [25].

The MICS has evolved into MedRadio, expanding the spectrum available for medical and neural implants [26]. However, since MedRadio transceivers restrict interference regulations, they are required to function with other medical implants in the same frequency bands. On the other side, the standard limits on power and bandwidth also further create challenges in applying RF communication for medical implants, capping their data rates at 267 kb/s [27]. The ultra-wideband (UWB) spectrum has raised strong interest since it supports higher rates, thus enabling reliable communication between a medical implant and an external device at lower costs and complexities. Besides, it enables low power, safety, and interference reduction. UWB has been studied extensively for different in-body communications applications [28]. Despite several advantages, applying UWB requires careful analysis of safe power limits according to specific absorption ratio (SAR) rules, and it has limits on propagation depths at higher frequencies of UWB.

Ensuring the security and privacy of patient data is critical, necessitating robust authentication and encryption methods to prevent unauthorized access. A critical drawback of RF for medical implants is that this technology remains vulnerable to hacking, underscoring the need for enhanced protection of RF links to secure patient information. On the other hand, in situations where multiple WBAN sensors are nearby within the hospital environment, RF signal interference can become a major issue, potentially affecting data integrity. For medical information and communication technology (ICT), data integrity is one of the crucial aspects [29]. Hence, risks associated with unauthorized access or intentional jamming of RF signals and interference issues must be carefully managed when RF communication is applied as an in-body carrier. Several methods have been intended to ensure authentication and prevent disruptions, such as cryptography [30]. Strong authentication can greatly reduce the risk of eavesdropping, and cryptographic keys based on physiological variations help protect data from unauthorized access. However, there is a trade-off, and it will probably create new issues: Using additional cryptography would consume more power for medical implants, and the inherent nature of the battery is limited energy and lifetime spans [6]. In other words, since medical implants are energy— and complexity—limited, implementing complex cryptographic mechanisms is not a viable solution in most cases.

2.2 Acoustics communication

Ultrasound refers to high-frequency acoustic waves (above 20 kHz) that typically are inaudible to humans [20]. Acoustics technology has been used since the 1940s to diagnose conditions such as heart issues and pregnancy. Nowadays, it is widely employed in numerous clinical applications, especially for diagnostic imaging. Due to lower tissue attenuation, acoustic waves can travel through biological tissues and have the potential to create wireless links with medical implants. Compared to RF, ultrasounds travel better through water, making them more suitable for data transmission in the human body, which is composed of water, that is, almost 70% [31]. Recent studies have shown the viability of acoustics communication in biological tissue; for instance, Weber et al. demonstrated that a 40 Kb/s data rate using only 100 μW of transmit power through 12 cm of castor oil can be attained for an implantable pressure sensor [32]. Later, Bos et al. reported that a data rate of 340 Kb/s was attained through beef samples using a small 2-mm transducer [33]. In Ref. [34], authors demonstrated an ultrasonic wideband system capable of 700 Kb/s transmission through a phantom at 10 cm, consuming around 40 μW while maintaining a low bit error rate. Higher data rates of acoustic communication can be attained with larger transducers, that is, greater than 1 cm [35], such as in [34, 36, 37], where data rates up to 30 Mb/s were demonstrated.

The main drawback is that acoustics waves cannot propagate well through the air medium because air is a poor conductor for ultrasound, so direct contact with the skin is required when employing ultrasound for wireless communication (zero air gap of transmitter distance). Piezoelectric transducers are mostly employed as the main transceivers for acoustic communication [3, 38], and a high data rate requires large-size transducers. However, large transducers may restrict their implementation in tiny medical implant cases [35]. Despite WBAN systems consuming very low power and acoustic communication using low energy, prolonged exposure is not recommended for safety concerns (i.e., tissue heating), which is not suitable for continuous monitoring [14]. During acoustic wave propagation, a portion of the energy is absorbed by the biological tissue and subsequently converted into heat, which can trigger a temperature increase. As wave intensity increases, the temperature rises. If it exceeds 38.5°C, it may lead to adverse biological effects, which can have a negative impact on the long-term reliability and sustainability of continuous operation [31].

2.3 Optical wireless communication (OWC)

OWC uses the entire light spectrum (including infrared, visible, and ultraviolet) for data transmission. OWC is becoming an important future technology, using light to carry data, similar to fiber optics. OWC offers various advantages, including resistance to channel interference [3], low complexity, and energy efficiency for low- to moderate-rate applications. Furthermore, it promotes safe, secure, and private connections. In the context of in-body carriers, OWC performs better than ultrasound, which can work in both contact and non-contact modes depending on how close the transmitter is to the human body. The inherent characteristic of OWC is very secure, as the communication coverage is limited, which practically protects against various eavesdropping and attacks, where the driver in the transmitter part safely controls the optical power at a permissible intensity; this security characteristic might not always be present in RF communication systems, resulting in systems that are vulnerable to attacks. The visible light spectrum can be used for data transmission to medical implants, as demonstrated in Refs. [39, 40], which employs white LED transmitters. However, the penetration capability of visible light is limited due to the significant absorption by the water contents of the human body. NIR light appears to the best approach as it works well in biological tissues due to minimal absorption and scattering. In the context of clinical applications, NIR light has been widely employed in diagnostics, low-level therapy applications, and pulse oximetry to monitor blood pressure [14].

By seeing the benefits of light (e.g., enhanced security and privacy, as optical links are immune to remote access and interference), OWC can be seen as attractive candidate for wireless communication in medical implants. The IEEE 802.15.7 standard, which was introduced in 2011, was one of the first for OWC, focusing on the visible light spectrum operating in free-space. Meanwhile, infrared wireless communication was first standardized by the Infrared Data Association (IrDA). The TG7r1, a target group under IrDA, was later formed to expand this standard, aiming to cover wavelengths from 190 to 10,000 nm, including ultraviolet, visible, and infrared light. We found that in the literature, only a few studies explored light specifically for in-body communication. On the other hand, there are no official standards for light communication for in-body communication, defining its spectrum, required bandwidth, minimum data rates, safety exposure, etc. For this reason, exploring this research area is a particularly new avenue; there are various open fascinating topics in this field. A recent study in our research group showed promising results in in-body communication using light: NIR light can be used for communication through biological tissues [14]. It was confirmed that NIR light could establish communication links with a data transmission speed of 10s kb/s and across tissue depths up to 40 mm.

In this chapter, we exploit OWC for in-body communication with the additional benefits of energy harvesting. We will explicitly show the feasibility of light in providing a communication link to adjust the implant operation wirelessly (forward biotelemetry) and harvest energy simultaneously. The proof-of-concept will be done using a testbed developed from commercially available equipment purchased from various vendors. Using light, medical implants are envisioned to be externally controlled effectively and securely, enabling medical staff to do several crucial tasks, for example, modify configurations or operating modes, manage system functions, and incorporate new medical profiles.

3. Wireless power transfer (WPT) method for medical implants

Energy sources for powering the medical implants' battery can be divided into two categories: internal sources, which are generated from the human body itself (e.g., human movement, human heat temperature), and external sources, which are harvested from outside the body [41]. External energy sources, such as RF, ultrasound, and light, are often employed for this purpose [20]. This approach is commonly called WPT; it enables the supply of energy without the need for lead wires for implanted device operation. Compared to internal energy sources, the WPT method facilitates more control of the power provided to the medical implants and exhibits superior power capabilities [42].

In this context, power can be sent wirelessly across the biological tissue. The main types of transmitting ends for WPT are electromagnetic antennas, ultrasound transducers, and light sources. These transmitter types send power to implanted devices equipped with receiving antennas, acoustic transducers, or PV cells [18]. As the demand for communication with implanted devices grows, continuous energy transfer to the implanted devices from external sources is becoming essential [43]. The following sections will discuss the benefits and drawbacks of WPT technology, focusing on RF, ultrasound, and light.

3.1 Radio frequency (RF) power transfer

In general, RF power transfer has gained importance for its flexibility in powering most electronic devices nowadays, such as smartphones, wearable devices, and others [44]. There are currently two RF power transfer methods: far-field and near-field [20]; differences in magnetic flux density between these two types present unique challenges. In the near field, designers struggle with optimal impedance matching and precise antenna alignment, while in the far field, the lower magnetic flux density makes it hard to collect enough energy to power the circuit [8].

RF power transfer shows promise for utilization in medical devices, especially implanted devices (e.g., pacemakers, brain implants, defibrillators), as it is anticipated to prolong the battery lifetime [45]. RF power transfer is widely used in therapeutic devices that stimulate different parts of the body for treatment purposes [20]. This technique employs a coil antenna to transmit electromagnetic waves, while a second coil antenna receives these waves and converts them into electrical current. Ferrite shielding can enhance the directionality of the electromagnetic field. However, high exposure to RF power transfer may generate excess heat at the medical implant site, posing a potential safety risk [45]. Another major challenge is energy loss through biological tissue, which can be reduced at sub-GHz frequencies [8].

Factors such as the separation between the devices, surrounding conditions, and the operational frequency influence the efficiency of this power transmission. RF power transfer does not need an exact alignment or proximity between the transmitter and receiver, as the emission coverage is spread [8].

3.2 Ultrasound power transfer

Ultrasound is defined as pulses of mechanical energy that propagate through various media and exhibit behaviors such as reflection and refraction upon interaction with objects [20]. When interacting with biological tissues, these waves demonstrate effective transmission. Historically, ultrasound technologies have been employed primarily for biological imaging applications, and there has been a growing interest in WPT studies for implanted devices nowadays [18]. This section briefly highlights ultrasound technology for acoustic power transmission, which facilitates the WPT function across short distances.

The working principle of ultrasound power transfer entails the conversion of electrical energy into ultrasound waves that are subsequently directed toward a receiver capable of converting them back into electrical energy for use in implanted devices. Notably, a receiver that is able to receive the transmitted ultrasound signals is needed [20]. Piezoelectric materials are predominantly utilized due to their ability to transduce ultrasound energy into electrical energy effectively *via* a vibrating resonant mode. The attenuation of ultrasound power within biological tissues is significantly less than that experienced by RF power transfer, allowing ultrasound to penetrate deeper into tissues with less energy dissipation. Ultrasonic power transfer can achieve a power transfer efficiency of over twenty-five percent [46] and penetrate up to 100 mm deep in biological tissue [47]. Furthermore, ultrasound power transfer is frequently employed in millimeter-sized medical implants (up to 500 mm^2) [46], as the dimensions of the receiver can be considerably smaller than those of the inductive coil [45], making it ideal for directional power transfer.

Although ultrasound technology excels at delivering energy into biological tissues, it faces challenges where the transmitter must be physically aligned with the skin, necessitating a mechanical connection between them [48]. Furthermore, impedance mismatches between different material types also become a drawback; when ultrasound waves hit boundaries between materials with different acoustic properties, much of the energy is reflected instead of transmitted. Therefore, ultrasound transmitters must maintain direct contact with the skin (zero gap), usually with the help of an acoustic coupling medium like gel. Additionally, ultrasound is less effective near the lungs and in areas surrounded by bone, such as the brain, requiring specially designed focused ultrasound transmitters for effective energy delivery [18]. Accordingly, ultrasound power is not possible to deliver power through the skull [49]. Ultrasound works at much lower frequencies in tissue, leading to shorter wavelengths compared to RF waves [50]. On the other hand, certain patients report experiencing pain and discomfort from noise during the utilization of this technology [43]. Addressing these concerns is essential for the widespread adoption of the technology.

3.3 Optical power transfer

Using light to generate energy has been common for powering portable devices [9]; nowadays, there are various commercial products that utilize light for power charging. In the context of healthcare applications, light can also be employed to charge implanted devices; it can be an uncontrollable power source (e.g., sunlight) or a controllable source (e.g., NIR LED emitting NIR light) that acts as a transmitter. The receiver typically uses a PV cell to receive the transmitted light and convert it into electricity. Examples of charging implanted devices using sunlight can be found in Refs. [51–55], demonstrating that implanted PV cells can generate reasonable output voltage in an open circuit (V_{OC}) from the sunlight penetrating skin. However, seasonal changes and geographical location also impact sunlight availability, leading to inconsistent power generation of PV cells, which contribute to efficiency reduction [56]. Sunlight penetration through individuals with darker skin tones is limited due to higher melanin levels, absorbing more light [54]. On the other hand, the thickness and type of biological tissues affect light penetration. Our previous study revealed that thicker tissue with more fat content allows less light to reach deeper [57]; in the context of energy harvesting using PV cells for deeper-implanted devices, using sunlight sources is not viable.

A controllable light source can be an effective solution for a deep-implanted device [50]. This method, commonly called optical power transfer, employs an LED or laser as a transmitter device. On the receiver side, the PV cell is employed to receive optical power from an external source [58]. Optical power transfer avoids electromagnetic interference (EMI), offering an advantage over RF power transfer [59, 60]. Another advantage is secure power transfer to implanted devices can be guaranteed. In our previous work [58], we demonstrated optical power transfer using 810 nm NIR light across *ex vivo* porcine samples with a monocrystalline silicon PV cell as a receiver. The results showed that a reasonable V_{OC} can be attained through 30 mm thickness.

The efficiency of optical power transfer depends on factors such as PV cell performance, the PV cell's wavelength coverage, the scattering and absorption coefficient of the tissue, and the alignment between the transmitter and receiver. Despite commercial PV technology having matured, PV cells for subcutaneous applications are still challenging, for instance, issues with compatibility with different tissue shapes. Commercial PV cells are typically designated for wearable devices, and they are optimized to absorb a wide spectrum of sunlight. Besides, PV cells can be also optimized for artificial (indoor) lights. They are not specifically designed to capture narrowband NIR light, especially Si-based PV cells, so the efficiency in harvesting energy from NIR light is relatively low. Efforts are underway to enhance the efficiency of NIR light absorption, for instance, Ref. [61] proposed organic PV cells and perovskite-silicon tandem cells to better harness NIR light. However, specialized NIR PV cells are still under exploration, which are on a laboratory scale (research stage), and have not yet been commercialized widely. Using standard PV cells for experiments on optical power transfer across biological tissue is still possible, but the harvested energy might require a reasonable exposure time. Obviously, there are open research areas in advancing PV cells for implanted device applications, that is, enhancing efficient power conversion covering all spectrums of light, designing biocompatibility-based flexible PV cells, and developing friendly materials for PV cells.

Similar to RF and ultrasound power transfer, optical power transfer can also enable data transfer by modulating the light beam's intensity, frequency, or phase to encode data. In this chapter, we exploit light to transfer data to implanted devices, and simultaneously, the commercial tiny PV cell is used to harvest energy from the transmitted light. Using light, it is anticipated that medical implants can be externally recharged while simultaneously facilitating data transmission. This method is expected to enhance the longevity of implanted devices for patients.

4. Joint data transmission: Energy harvesting

4.1 Concepts

This subsection explains the concept of joint data transmission and energy harvesting for implanted devices. As described in Section 2, a light source is able to establish a wireless link and transmit power wirelessly through biological tissue by allowing the data receiver to capture optical signals and the energy harvester to convert light passing through the tissue into considerable power. This method provides two benefits in a row: highly secure communication for passive-located implanted devices (for instance, pacemakers, cochlear implants, defibrillators, insulin pumps) in the future and reliable power charging.

Communication with the implanted device can be divided into two main ways: from inside the body (in-body) to outside (out-body) or vice versa, and communication within the in-body devices, as shown in **Figure 1(a)**. The details of both scenarios are further described in Refs. [24, 57]. This chapter focuses on the first approach, that is, the optical link between the in-body and out-body devices. Here, we consider transmitting data from the out-body to the in-body devices to represent "forward biotelemetry." The concept aims to enable the adjustment of operational settings and updating of patient profiles while powering implantable devices. In addition to signals related to device settings, as mentioned before (e.g., user profiles), the transmitted signals to the implanted devices could include signals controlling actuators, such as actions for an insulin pump. The big picture of system architecture, as shown in **Figure 1(b)**, includes a digital signal processing (DSP) unit transceiver, an optical front-end transceiver, an optical transmitter, and two types of receivers: a photodetector and a PV cell.

The transmitter device can utilize an NIR LED or a vertical-cavity surface-emitting laser (VCSEL) diode. Data from an out-body device is processed by the DSP, transmitted optically by the NIR LED, and received through biological tissue by the photodetector. The DSP subsystem then demodulates the data and sends it to the stimulus device.

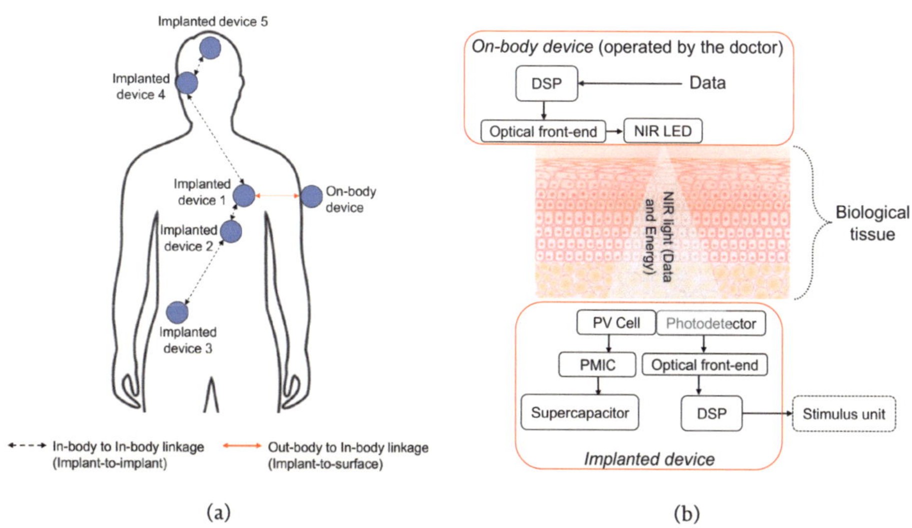

Figure 1.
(a) Generic architecture of WBAN containing the linkage between on-body and in-body devices, reproduced from Ref. [57] with permission; (b) used architecture (i.e., implanted device).

NIR LED is considered in this case as it provides wider coverage and thus performs better in terms of misalignment than VCSEL. On the receiver side, a photodetector placed on the opposite of the NIR LED converts the light signal into electrical signals for demodulation and command detection; at the same time, a PV cell placed near the photodetector captures energy from the optical signal. As aforementioned in Section 3.3, most commercial PV cells nowadays are usually engineered to work effectively with a broad light spectrum, that is, sunlight or artificial light [62]. In our case, we use NIR light at a specific wavelength (i.e., 810 nm), which is a much narrower part of the spectrum. Although NIR light can penetrate biological tissue, part of the light is absorbed and scattered; thus, the optical power experiences attenuation. This is a challenging task, as harvesting energy quickly is not possible. For this reason, we need to harvest for a considerable period of time in order to gather enough energy and store it in a supercapacitor or other energy sources (e.g., rechargeable batteries). The role of the energy harvesting unit, that is, power management integrated circuit (PMIC), is very important here; it bridges a PV cell and a supercapacitor. The implanted device could then utilize the stored energy in the supercapacitor during its operation. The storage can use either a rechargeable battery or a supercapacitor. In the subsequent subchapter, we explain the demonstration of the proposed concept.

4.2 Proof-of-concept demonstrations

Developing and testing new medical devices involves multiple stages. It begins with preclinical studies, often conducted in a test bed (involving *ex vivo* and *in vivo* tests), to obtain basic information about the feasibility. It then progresses to clinical trials in humans. This work considers experiments in the test bed since it is a very initial stage, in which we just show the experiments' initial results that prove the concept's viability. A test bed is built by strictly referring to the big picture of the system, as depicted in **Figure 1(b)**, using commercial equipment purchased from various vendors. The diagram block of the test bed and a photograph of the experimental laboratory setting are shown in **Figure 2(a)** and **(b)**, respectively.

The test bed includes optical transmitters, receivers, and propagation medium, which are detailed in **Table 1**. It employs two National Instrument Universal Software Radio Peripheral (NI USRP) devices and personal computers; one serves as the host, and the other is the client. The DSP is built by GNU Radio on a Linux operating system. This chapter does not discuss DSP system architecture, and the details will appear in subsequent publications. A NIR LED serves as the transmitter, operating in "external modulation" mode controlled by an LED driver. This mode allows a connection between the NIR LED and NI USRP. In this experiment, we consider an optical phantom to mimic the human soft tissue, where the recipe is carefully prepared based on Ref. [63]; it is a TuO_2 powder-based recipe called the *polyvinyl chloride plastisol* (PVCP) mixture. Its optical properties correspond to those available in average human soft tissue [64]. The phantom is 40 mm thick (**Figure 2a**); we chose this thickness because it is close to the maximum range possible with optical OWC, as demonstrated in Ref. [14]. The phantom serves as the propagation medium and is placed between the NIR LED transmitter and receivers (a photodetector and a PV cell). The phantom performs better than actual porcine samples over an extended period of measurements as it remains stable and does not change its optical characteristics too much, while the porcine samples can. Actual porcine samples are not suitable for long-term experiments, as carried out in this study, where considerable time is needed during data transmission and wireless charging.

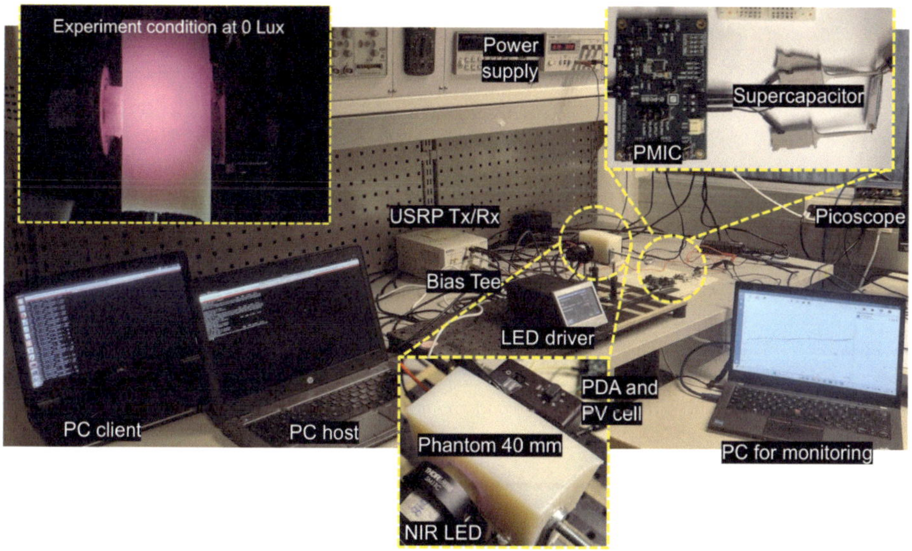

(a)

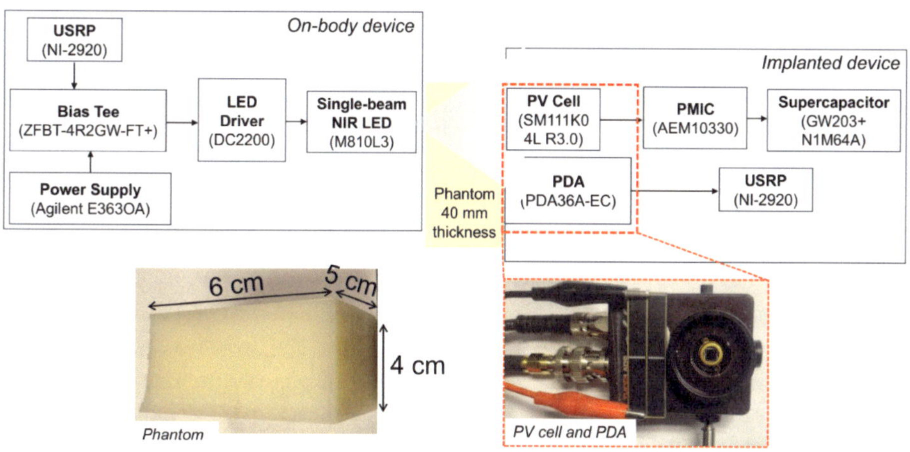

(b)

Figure 2.
(a) A diagram block of the test bed; (b) a photograph of the experimental setup on optical wireless power and information transfer across the 40-mm-thick phantom using NIR light 810 nm.

The photodetector amplifier (PDA) used to receive data was set to 0 dB gain with no lenses. A 25% efficiency monocrystalline silicon PV cell used to harvest NIR energy was placed adjacent to the PDA (**Figure 2a**). Given that there is a challenge in using a PV cell as an energy harvester in deep biological tissue, we prioritize a PV cell to get a considerable light intensity level. For this reason, the NIR LED beam is positioned to have a strict directional path with the PV cell to ensure it receives sufficient optical power, as shown in **Figure 2(a)**. By this setting, the PDA will receive a smaller portion of light to receive data, but it is still possible to capture enough optical power. This study considers a supercapacitor to store harvested energy from NIR light.

Units	Components	Description	Vendors	Source
Propagation medium	Phantom 40 mm thickness	PVCP-based material	Recipes from Refs. [63, 64]	
Transmitter	NIR LED	M810L3 (375 mW maximum optical power, λ = 810 nm)	Thorlabs	[65]
	DSP at the transmitter and receiver	USRP NI-2920	(National Instruments)	[66]
	Bias tee	ZFBT-4R2GW-FT+	Mini-Circuits	[67]
	LED driver	DC2200	Thorlabs	[68]
Receiver	Photodetector	PDA36A-EC	Thorlabs	[69]
	PV cell	PV cell SM111K04L R3.0	IXOLAR	[70]
	PMIC development kit	AEM10330	E-Peas	[71]
	Supercapacitor	GW203+ N1M64A	CAP-XX	[72]

Table 1.
Key components of the test bed.

The setting of PMIC AEM10330 in this experiment is as follows: a wide swing voltage (1–4.5 V), 80% maximum power point tracking (MPPT), a 341 ms sampling duration, a 21.8 s sampling period, and the balance (BAL) option, as the supercapacitor was a double-cell type. Then, the harvested energy can be calculated using Eq. (1) [73]; it is approximately 5.005 Joules.

$$E_{store} = \frac{1}{2} \times C \times \left(V_{full}^2 - V_{\min}^2 \right). \tag{1}$$

where C is the supercapacitor capacitance (i.e., 0.52F), V_{full} is the final voltage (i.e., 4.5 V), and $V_{\min}$ is the initial voltage (i.e., 1 V).

The NIR LED is a nonlinear source, so in order for the modulated signal to be positioned at the linear region of NIR LED, a commercial Bias tee was employed; the modulated signal from the NI USRP is biased through Bias tee with external power supply as an input. The LED driver's datasheet suggests that in 'external modulation' mode, per 1 V input given by the external power supply, 400 mA of current is generated for the LED. In this case, the NIR LED has 500 mA of maximum driving current. Applying 2 V would generate 800 mA, which damages the NIR LED. For this experiment, the external power supply was set to 0.2 V, delivering approximately only 80 mA of current (the power density using this driving current was 66.5 mW/cm^2), which raises safety considerations for NIR LED 810 nm and fills the maximum permissible exposure (MPE) aspect, that is, below 100 mW/cm^2, which is irradiance of NIR light at 810 nm typically used in low-light level therapy (LLLT) [74].

We then demonstrate a joint data transfer and energy harvesting across phantom in a single direction after the test bed is ready. Our experiment here does not consider misalignment issues. The directional nature of light means that any misalignment between the transmitter and receiver can lead to a loss of optical signal [75]. OWC is sensitive to misalignment, which is probably problematic in a dynamic environment where the relative positions of implanted devices may change over time. Our previous study revealed that misalignment between out-body and in-body devices affects the received power loss [75]. Since this study is an initial investigation, we will consider misalignments (lateral or angular) and pointing errors in future studies.

We chose the external modulation mode provided by the LED driver. Afterwards, we transferred GMSK-modulated data from the PC host to the client using UDP protocol and the tunnel.py library across a phantom medium. We considered a 26 MB high-definition image file in TIFF format to represent commands in the forward biotelemetry case. During the data transfer, the supercapacitor's charging progress is observed in real time using an oscilloscope (Picoscope 3000 series) connected to a personal computer. We conducted the data transmission three times in a row. The supercapacitor was initially set to approximately 1 V, and we observed its charging behavior during data transmission starting from this voltage level. As shown in

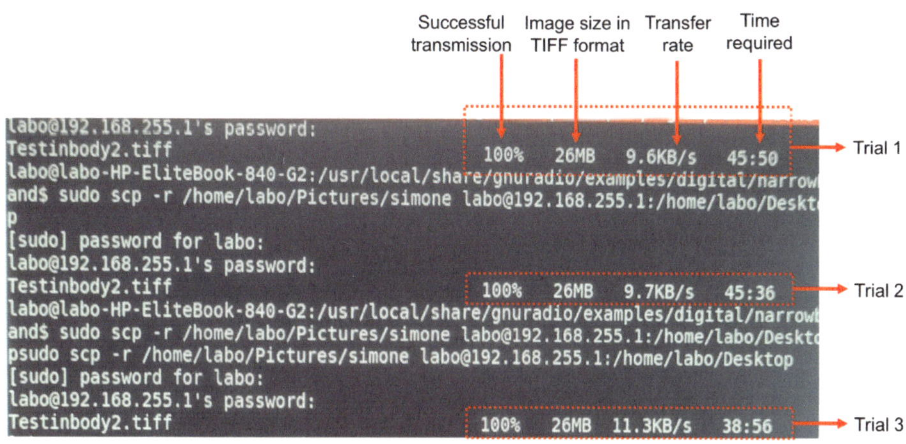

(a)

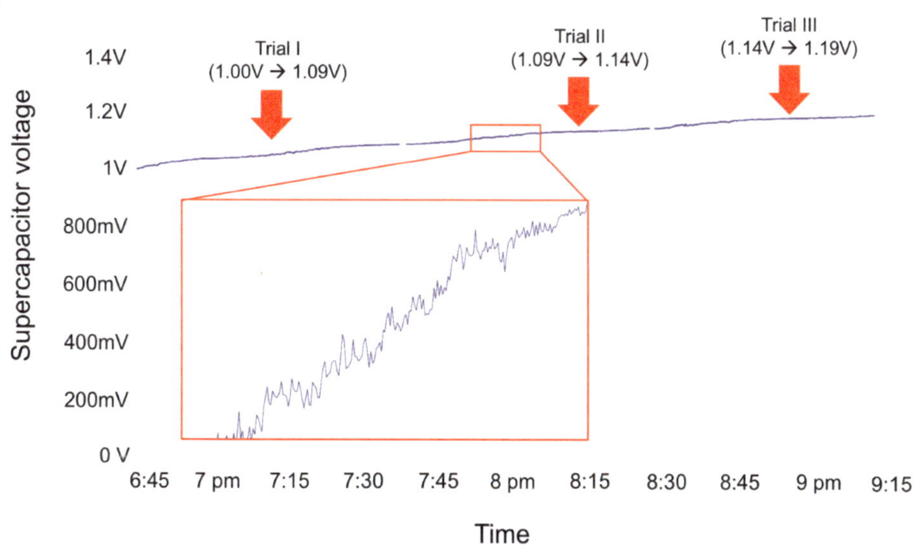

(b)

Figure 3.
Demonstration of joint data transmission and energy harvesting using NIR LED as a transmitter and PV cell as receiver across a 40-mm-thick phantom: (a) transmission report and (b) data log of the charging supercapacitor during the data transmission.

Figure 3(a), the transfer rates for the first, second, and third attempts to transmit a 44.1 MB image on 100% successful are 9.6, 9.7, and 11.3 kb/s, respectively.

The results showed the supercapacitor charges during the three attempts of file transmissions. **Figure 3** shows the charging characteristics during these transmissions through the phantom. The V_{OC} and short-circuit current (I_{SC}) of the PV cell, when it was illuminated with NIR, LED using 'external modulation' were 0.79 V and 0.07 mA, respectively. As depicted in **Figure 3**, the voltage increases slightly from 1.00 to 1.09 V, 1.09 to 1.14 V, and 1.14 to 1.19 V for the first, second, and third attempts, respectively. The increasing voltages from the first, second, and third tests are 0.09, 0.05, and 0.05 V, respectively. The corresponding time required for complete image transmission in these trials was 45 mins 50 sec, 45 mins 36 sec, and 38 mins 56 secs, respectively (based on the successful transfer report on the PC host). As a note, a total energy of 5.005 Joules can be achieved when the supercapacitor is fully charged to its maximum voltage of 4.5 V, with the measurements commencing from an initial voltage of 1 V. However, in the present demonstration, the supercapacitor does not attain full charge, achieving only 1.9 V following three attempts of image transmission. Typical commands that are sent to the medical device for the activation are short, for example, within an hour.

As expected, the PV cell received less power due to scattering and absorption when the signal passed through a 40-mm-thick phantom, resulting in slower charging times. Despite the fact that we need hours to charge the supercapacitor, the solution is practically acceptable. The charging could take place, for instance, when the patient is sleeping. The gradually accumulated energy can contribute to prolonging the operational lifespan of the implanted device. Typically, the current consumption of such implants is relatively low, ranging from micro-watts to milliwatts [62, 76, 77]. Accordingly, even a modest amount of stored energy is anticipated to sustain the device's operation for the next hours or a few days. A new investigation in our group revealed that when the supercapacitor is fully charged (at 4.5 V, as the maximum voltage), it is capable of powering implanted devices (e.g., pacemakers) for more than 12 hours, assuming that the current consumption is 50 μA. The proposed method is secure and safe, leveraging key advantages of the optical system.

The primary objective of this study is to exploit the use of NIR LEDs for communication purposes through biological tissue while simultaneously facilitating energy harvesting during data transmission. In this idea, the PV cell harvests the energy from NIR light emission, which carries data. It is acknowledged that the amount of energy harvested in this context is relatively small since the energy is transmitted periodically based on the control demands of the implanted device (i.e., only when the doctor is operating implanted devices within the patient wirelessly); this scheme is called a periodic charge scheme. In scenarios where energy harvesting is given greater precedence, continuous power transmission with a permissible illumination level of NIR light can be a solution. In this case, the optical signal contains solely a direct current component (stable illumination over time). This approach can be called a full-charge scheme [44], which is anticipated to charge the supercapacitor faster due to the higher illumination of the NIR LED that transmits the power.

5. Conclusion and future work

This chapter presents the concept and practical implementation of joint energy harvesting and data transfer for medical implants exploiting light. This chapter also enriches the contents by discussing the advantages and disadvantages of methods for data transmission and WPT with medical implants using three distinct technologies, that is, light, RF, and acoustic, each of which has some advantages and disadvantages. Light is anticipated to have better security and safety for data transmission across biological tissue, and it is EMI-free. Therefore, safe data transmission and energy charging to the implanted devices are believed to be guaranteed.

This preliminary study's proof of concept was demonstrated in a test bed using NIR light. The key components of the experimental system are a NIR LED, PV cell, and tissue-mimicking optical phantom as transmitter, receiver, and propagation medium, respectively. Biological tissues scatter and absorb light, considerably deteriorating the optical signal as it travels through the body. NIR light 810 nm is used because it particularly propagates well in biological tissues. We have demonstrated that charging supercapacitors concurrently with data transmission is feasible. Still, this process is notably time-consuming as the rate of voltage increase in the supercapacitor tends to be very slow, that is, approximately 0.05 V. Wireless charging can be conducted during the patient's sleep, thereby minimizing any disruption to their daily activities. Despite the small increase in voltage of the supercapacitor, the further harvested energy is envisioned to assist in prolonging the implanted device's lifespan, which typically consumes very low-to-moderate power. Even a small amount of stored energy can sustain the device for hours. We could also charge rechargeable batteries; they could store more energy but also need more time to charge.

To improve the efficiency of energy harvesting from NIR LEDs, NIR-optimized PV cells in conjunction with the broadband NIR LED transmitter can be considered. This would thereby ensure a reliable and adequate power supply for implanted devices due to the spectral match between the wavelength coverage of the PV cell and NIR light emission. Future research endeavors will also focus on the exploration of specialized PV cells designed for subcutaneous applications, that is, thin, flexible, and biocompatible. There is also an interesting topic in the development of a novel wireless communication protocol specifically designed for this application aimed at optimizing both energy harvesting efficiency and data transmission capabilities. Furthermore, this protocol should exhibit resilience to noise, facilitate higher data transmission rates, and support both continuous and pulsed power transmissions.

Acknowledgements

This work was funded by the University of Oulu's Infotech (CWC-NS 2406124). The authors would like to acknowledge the University of Oulu and the Research Council of the Finland 6G Flagship Programme (Grant Number: 369116).

Conflict of interest

The authors declare no conflict of interest.

Author details

Syifaul Fuada[1*], Mariella Särestöniemi[1,2,3] and Marcos Katz[1]

1 Centre for Wireless Communications, Faculty of Information Technology and Electrical Engineering, University of Oulu, Oulu, Finland

2 Health Sciences and Technology, Faculty of Medicine, University of Oulu, Oulu, Finland

3 InfoTech Oulu, Finland

*Address all correspondence to: syifaul.fuada@oulu.fi

References

[1] Sohail A, Ali A, Shaukat H, Bhatti FM, Ali S, Kouritem SA, et al. Integrating self-powered medical devices with advanced energy harvesting: A review. Energy Strategy Reviews. 2024;**52**:101328

[2] Busnatu Ștefan S, Niculescu AG, Bolocan A, Andronic O, Pantea Stoian AM, Scafa-Udriște A, et al. A review of digital health and biotelemetry: Modern approaches towards personalized medicine and remote health assessment. Journal of Personalized Medicine. 2022;**12**(10):1656

[3] Teshome AK, Kibret B, Lai DTH. A review of implant communication technology in WBAN: Progress and challenges. IEEE Reviews in Biomedical Engineering. 2019;**12**:88-99

[4] Kiourti A, Nikita KS. A review of In-body biotelemetry devices: Implantables, ingestibles, and injectables. IEEE Transactions on Biomedical Engineering. 2017;**64**(7):1422-1430

[5] Vizziello A, Magarini M, Savazzi P, Galluccio L. Intra-body communications for nervous system applications: Current technologies and future directions. Computer Networks. 2023;**227**:109718

[6] Fuada S, Ma G, Katz M. Global growth and trends of In-body communication research—Insight from bibliometric analysis. International Journal of Online and Biomedical Engineering (iJOE). 2024;**20**(01):128-149

[7] Owida HA, Al-Nabulsi JI, Turab NM, Alnaimat F, Rababah H, Shakour MY. Autocharging techniques for implantable medical applications. International Journal of Biomaterials. 2021;**2021**(1):6074657

[8] Gao Z, Zhou Y, Zhang J, Foroughi J, Peng S, Baughman RH, et al. Advanced energy harvesters and energy storage for powering wearable and implantable medical devices. Advanced Materials. 2024;**36**:2404492. DOI: 10.1002/adma.202404492

[9] Zhao J, Ghannam R, Htet KO, Liu Y, Kay LM, Roy VAL, et al. Self-powered implantable medical devices. Photovoltaic Energy Harvesting Review. Advanced Healthcare Materials. 2020;**9**(17):2000779

[10] Essa A, Almajali E, Mahmoud S, Amaya RE, Alja'Afreh SS, Ikram M. Wireless power transfer for implantable medical devices: Impact of implantable antennas on energy harvesting. IEEE Open Journal of Antennas and Propagation. 2024;**5**(3):739-758

[11] Fuada S, Särestöniemi M, Katz M. Analyzing the trends and global growth of energy harvesting for implantable medical devices (IMDs) research—A bibliometric approach. International Journal of Online and Biomedical Engineering (iJOE). 2024;**20**(03):115-135

[12] Abouelmehdi K, Beni-Hessane A, Khaloufi H. Big healthcare data: Preserving security and privacy. Journal of Big Data. 2018;**5**(1):1

[13] Balancing Act: Ensuring Patient Privacy in Health Information Exchange [Internet]. New York, USA: Reveal HealthTech.; 2024. Available from: https://revealhealthtech.com/blogs/balancing-act-ensuring-patient-privacy-in-health-information-exchange/ [Accessed: November 17, 2024]

[14] Ahmed I, Halder S, Bykov A, Popov A, Meglinski IV, Katz M.

In-body communications exploiting light: A proof-of-concept study using ex vivo tissue samples. IEEE Access. 2020;**8**:190378-190389

[15] Halder S, Säreströniemi M, Ahmed I, Katz M. Providing connectivity to implanted electronics devices: Experimental results on optical communications over biological tissues with comparisons against UWB. In: Alam MM, Hämäläinen M, Mucchi L, Niazi IK, Le Moullec Y, editors. Body Area Networks Smart IoT and Big Data for Intelligent Health, Lecture Notes of the Institute for Computer Sciences, Social Informatics and Telecommunications Engineering. Cham: Springer International Publishing; 2020. pp. 3-17

[16] Ash C, Dubec M, Donne K, Bashford T. Effect of wavelength and beam width on penetration in light-tissue interaction using computational methods. Lasers in Medical Science. 2017;**32**(8):1909-1918

[17] Upputuri PK, Pramanik M. Photoacoustic imaging in the second near-infrared window: A review. Journal of Biomedical Optics. 2019;**24**(4):040901

[18] Singer A, Robinson JT. Wireless power delivery techniques for miniature implantable bioelectronics. Advanced Healthcare Materials. 2021;**10**(17):2100664

[19] Zheng G, Shankaran R, Orgun MA, Qiao L, Saleem K. Ideas and challenges for securing wireless implantable medical devices: A review. IEEE Sensors Journal. 2017;**17**(3):562-576

[20] Park W, Lee J, Chung WG, Jeong I, Kim E, Kwon YW, et al. Recent advances in wireless energy transfer technologies for body-interfaced electronics. Nano Energy. 2024;**124**:109496

[21] Magno M, Spagnol C, Benini L, Popovici E. A low power wireless node for contact and contactless heart monitoring. Microelectronics Journal. 2014;**45**(12):1656-1664

[22] Singh D, Eleonye T, Surazynski L, Ferdinando H, Kumar A, Joshi HD, et al. Preliminary studies on mm-wave radar for vital sign monitoring of driver in vehicular environment. In: Säreströniemi M, Keikhosrokiani P, Singh D, Harjula E, Tiulpin A, Jansson M, et al., editors. Digital Health and Wireless Solutions. Cham: Springer Nature Switzerland; 2024. pp. 480-493

[23] Khan MB, Rehman M, Mustafa A, Shah RA, Yang X. Intelligent non-contact sensing for connected health using software defined radio technology. Electronics. 2021;**10**(13):1558

[24] Mohamed M, Maiseli BJ, Ai Y, Mkocha K, Al-Saman A. In-body sensor communication: Trends and challenges. IEEE Electromagnetic Compatibility Magazine. 2021;**10**(2):47-52

[25] Sayrafian-Pour K, Yang WB, Hagedorn J, Terrill J, Yekeh Yazdandoost K, Hamaguchi K. Channel models for medical implant communication. International Journal of Wireless Information Networks. 2010;**17**(3):105-112

[26] Chow EY, Morris MM, Irazoqui PP. Implantable RF medical devices: The benefits of high-speed communication and much greater communication distances in biomedical applications. IEEE Microwave Magazine. 2013;**14**(4):64-73

[27] Koprowski R. Overview of technical solutions and assessment of clinical usefulness of capsule endoscopy. BioMedical Engineering OnLine. 2015;**14**(1):111

[28] Särestöniemi M, Niemelä V, Hämäläinen M, Iinatti J, Keränen N, Jämsä T, et al. Receiver performance evaluation on IEEE 802.15.6 based WBAN for monitoring Parkinson's disease. In: 2014 Eighth International Symposium on Medical Information and Communication Technology (ISMICT) [Internet]. Firenze, Italy: IEEE; 2014. pp. 1-5. Available from: https://ieeexplore.ieee.org/document/6825220 [Accessed: October 25, 2024]

[29] Mucchi L, Jayousi S, Martinelli A, Caputo S, Marcocci P. An overview of security threats, solutions and challenges in WBANs for healthcare. In: 2019 13th International Symposium on Medical Information and Communication Technology (ISMICT) [Internet]. Oslo, Norway: IEEE; 2019. pp. 1-6. Available from: https://ieeexplore.ieee.org/abstract/document/8743798 [Accessed: October 21, 2023]

[30] Ankarali ZE, Abbasi QH, Demir AF, Serpedin E, Qaraqe K, Arslan H. A comparative review on the wireless implantable medical devices privacy and security. In: 2014 Fourth International Conference on Wireless Mobile Communication and Healthcare - Transforming Healthcare through Innovations in Mobile and Wireless Technologies (MOBIHEALTH) [Internet]. Athens, Greece: IEEE; 2014. pp. 246-249. Available from: https://ieeexplore.ieee.org/document/7015957 [Accessed: October 25, 2024]

[31] Galluccio L, Melodia T, Palazzo S, Santagati GE. Challenges and implications of using ultrasonic communications in intra-body area networks. In: 2012 Ninth Annual Conference on Wireless on-Demand Network Systems and Services (WONS) [Internet]. Courmayeur, Italy: IEEE; 2012. pp. 182-189. Available from: https://ieeexplore.ieee.org/document/6152227 [Accessed: October 25, 2024]

[32] Weber MJ, Yoshihara Y, Sawaby A, Charthad J, Chang TC, Garland R, et al. A high-precision 36 mm3 programmable implantable pressure sensor with fully ultrasonic power-up and data link. In: 2017 Symposium on VLSI Circuits [Internet]. Kyoto, Japan: IEEE; 2017. pp. C104-C105. Available from: https://ieeexplore.ieee.org/document/8008564 [Accessed: October 25, 2024]

[33] Bos T, Dehaene W, Verhelst M. Ultrasound In-body communication with OFDM through multipath realistic channels. In: 2019 IEEE Biomedical Circuits and Systems Conference (BioCAS) [Internet]. Nara, Japan: IEEE; 2019. pp. 1-4. Available from: https://ieeexplore.ieee.org/document/8918755 [Accessed: October 25, 2024]

[34] Santagati GE, Melodia T. Experimental evaluation of impulsive ultrasonic intra-body Communications for Implantable Biomedical Devices. IEEE Transactions on Mobile Computing. 2017;**16**(2):367-380

[35] Tabak G, Yang S, Miller RJ, Oelze ML, Singer AC. Video-capable ultrasonic wireless communications through biological tissues. IEEE Transactions on Ultrasonics, Ferroelectrics, and Frequency Control. 2021;**68**(3):664-674

[36] Singer A, Oelze M, Podkowa A. Mbps experimental acoustic through-tissue communications: MEAT-COMMS. In: 2016 IEEE 17th International Workshop on Signal Processing Advances in Wireless Communications (SPAWC) [Internet]. Edinburgh, UK: IEEE; 2016. pp. 1-4. Available from: https://ieeexplore.ieee.org/document/7536815 [Accessed: October 25, 2024]

[37] Demirors E, Alba G, Santagati GE, Melodia T. High data rate ultrasonic communications for wireless intra-body networks. In: 2016 IEEE International Symposium on Local and Metropolitan Area Networks (LANMAN) [Internet]. Rome, Italy: IEEE; 2016. pp. 1-6. Available from: https://ieeexplore.ieee.org/document/7548843 [Accessed: October 25, 2024]

[38] Haddad O, Khalighi MA. Enabling communication Technologies for Medical Wireless Body-Area Networks. In: 2019 Global LIFI Congress (GLC) [Internet]. Paris, France: IEEE; 2019. pp. 1-5. Available from: https://ieeexplore.ieee.org/document/8864122 [Accessed: October 23, 2024]

[39] Mosenia A, Jha NK. OpSecure: A secure unidirectional Optical Channel for implantable medical devices. IEEE Transactions on Multi-Scale Computing Systems. 2018;**4**(3):410-419

[40] Koschay M, Richter H, Statz M, Kober M, Puschmann J, Plocksties F, et al. Case-study on visible light communication for implant monitoring. In: 2022 IEEE Biomedical Circuits and Systems Conference (BioCAS) [Internet]. Taipei, Taiwan: IEEE; 2022. pp. 275-279. Available from: https://ieeexplore.ieee.org/document/9948625 [Accessed: January 4, 2024]

[41] Karimi MJ, Schmid A, Dehollain C. Wireless power and data transmission for implanted devices via inductive links: A systematic review. IEEE Sensors Journal. 2021;**21**(6):7145-7161

[42] Kim H, Rigo B, Wong G, Lee YJ, Yeo WH. Advances in wireless, batteryless, implantable electronics for real-time, continuous physiological monitoring. Nano-Micro Letters. 2023;**16**(1):52

[43] Amar AB, Kouki AB, Cao H. Power approaches for implantable medical devices. Sensors. 2015;**15**(11):28889-28914

[44] Alabsi A, Hawbani A, Wang X, Dubai AA, Hu J, Aziz SA, et al. Wireless power transfer technologies, applications, and future trends: A review. IEEE Transactions on Sustainable Computing. 2024:1-18. Available from: https://ieeexplore.ieee.org/document/10478158

[45] Nelson BD, Karipott SS, Wang Y, Ong KG. Wireless technologies for implantable devices. Sensors. 2020;**20**(16):4604

[46] Radziemski L, Makin IRS. In vivo demonstration of ultrasound power delivery to charge implanted medical devices via acute and survival porcine studies. Ultrasonics. 2016;**64**:1-9

[47] Donohoe M, Balasubramaniam S, Jennings B, Jornet JM. Powering In-body nanosensors with ultrasounds. IEEE Transactions on Nanotechnology. 2016;**15**(2):151-154

[48] Barbruni GL, Ros PM, Demarchi D, Carrara S, Ghezzi D. Miniaturised wireless power transfer Systems for Neurostimulation: A review. IEEE Transactions on Biomedical Circuits and Systems. 2020;**14**(6):1160-1178

[49] Agarwal K, Jegadeesan R, Guo YX, Thakor NV. Wireless power transfer strategies for implantable bioelectronics. IEEE Reviews in Biomedical Engineering. 2017;**10**:136-161

[50] Miziev S, Pawlak WA, Howard N. Comparative analysis of energy transfer mechanisms for neural implants. Frontiers in Neuroscience [Internet]. 2024;**17**(1320441):1-20. Available from: https://www.frontiersin.org/

journals/neuroscience/articles/10.3389/fnins.2023.1320441/full

[51] Haeberlin A, Zurbuchen A, Schaerer J, Wagner J, Walpen S, Huber C, et al. Successful pacing using a batteryless sunlight-powered pacemaker. EP Europace. 2014;**16**(10):1534-1539

[52] Haeberlin A, Zurbuchen A, Walpen S, Schaerer J, Niederhauser T, Huber C, et al. The first batteryless, solar-powered cardiac pacemaker. Heart Rhythm. 2015;**12**(6):1317-1323

[53] Bereuter L, Williner S, Pianezzi F, Bissig B, Buecheler S, Burger J, et al. Energy harvesting by subcutaneous solar cells: A long-term study on achievable energy output. Annals of Biomedical Engineering. 2017;**45**(5):1172-1180

[54] Tholl MV, Akarçay HG, Tanner H, Niederhauser T, Zurbuchen A, Frenz M, et al. Subdermal solar energy harvesting – A new way to power autonomous electric implants. Applied Energy. 2020;**269**:114948

[55] Wu T, Redouté JM, Yuce MR. A wireless implantable sensor design with subcutaneous energy harvesting for long-term IoT healthcare applications. IEEE Access. 2018;**6**:35801-35808

[56] Tholl MV, Zurbuchen A, Tanner H, Haeberlin A. Potential of subdermal solar energy harvesting for medical device applications based on worldwide meteorological data. Journal of Biomedical Optics. 2021;**26**(3):038002

[57] Fuada S, Särestöniemi M, Katz M, Soderi S, Hämäläinen M. Study on fat as the propagation medium in optical-based In-body communications. In: Särestöniemi M, Keikhosrokiani P, Singh D, Harjula E, Tiulpin A, Jansson M, et al., editors. Digital Health and Wireless Solutions. Cham: Springer Nature Switzerland; 2024. pp. 467-479

[58] Fuada S, Perera MAN, Särestöniemi M, Katz M. Optical wireless power transmission through biological tissue using commercial photovoltaic cells under 810 nm LEDs: Feasibility study. In: Särestöniemi M, Keikhosrokiani P, Singh D, Harjula E, Tiulpin A, Jansson M, et al., editors. Digital Health and Wireless Solutions. Cham: Springer Nature Switzerland; 2024. pp. 388-400

[59] Ding J, Liu W, Chih-Lin I, Zhang H, Mei H. Advanced progress of optical wireless technologies for power industry: An overview. Applied Sciences. 2020;**10**(18):6463

[60] Nguyen DH. Optical wireless power transfer for implanted and wearable devices. Sustainability. 2023;**15**(10):8146

[61] Jaysankar M, Filipič M, Zielinski B, Schmager R, Song W, Qiu W, et al. Perovskite–silicon tandem solar modules with optimised light harvesting. Energy & Environmental Science. 2018;**11**(6):1489-1498

[62] Wangatia LM, Yang S, Zabihi F, Zhu M, Ramakrishna S. Biomedical electronics powered by solar cells. Current Opinion in Biomedical Engineering. 2020;**13**:25-31

[63] Vogt WC, Jia C, Wear KA, Pfefer TJ. Biologically relevant photoacoustic imaging phantoms with tunable optical and acoustic properties. Journal of Biomedical Optics. 2016;**21**(10):101405

[64] Spirou GM, Oraevsky AA, Vitkin IA, Whelan WM. Optical and acoustic properties at 1064 nm of polyvinyl chloride-plastisol for use as a tissue phantom in biomedical optoacoustics. Physics in Medicine and Biology. 2005;**50**(14):N141

[65] Thorlabs. M810L3 810 nm, 325 mW (Min) Mounted LED, 500 mA [Internet].

Available from: https://www.thorlabs.com/thorproduct.cfm?partnumber=M810L3&pn=M810L3#4478 [Accessed: November 17, 2024]

[66] https://www.ni.com [Internet]. USRP-2920 specifications - NI. Available from: https://www.ni.com/docs/en-US/bundle/usrp-2920-specs/page/specs.html#ariaid-title2 [Accessed: November 17, 2024]

[67] BIAS TEE/SMA/RoHS. ZFBT-4R2GW-FT+ – Mini-Circuits [Internet]. Available from: https://www.minicircuits.com/WebStore/dashboard.html?model=ZFBT-4R2GW-FT%2B&srsltid=AfmBOorxk99tkZPbPEons3cfDDKL1YoxYy6u897aAldDZqmcxMxPlR_y [Accessed: November 17, 2024]

[68] Thorlabs. DC2200 high-power 1-channel LED driver with pulse modulation, 10.0 A Max, 50.0 V Max [Internet]. Available from: https://www.thorlabs.com [Accessed: November 17, 2024]

[69] Thorlabs. PDA36A-EC Si Switchable Gain Detector, 350-1100 nm, 10 MHz BW, 13 mm2, M4 Taps [Internet]. Available from: https://www.thorlabs.com [Accessed: November 17, 2024]

[70] Gen2. [Internet]. Available from: https://waf-e.dubudisk.com/anysolar.dubuplus.com/techsupport@anysolar.biz/O18Adzo/DubuDisk/www/Gen2/SM111K04L%20DATA%20SHEET%20202007.pdf [Accessed: November 17, 2024]

[71] AEM10330Solar energy harvesting [Internet]. E-peas. Available from: https://waf-e.dubudisk.com/anysolar.dubuplus.com/techsupport@anysolar.biz/O18Adzo/DubuDisk/www/Gen2/SM111K04L%20DATA%20SHEET%20202007.pdf [Accessed: November 17, 2024]

[72] Prismatic Ultra Thin Supercaps [Internet]. CAP-XX Designs & Manufactures the Thinnest, Highest Power Density Supercapacitors & Ultracapacitors. Available from: https://www.cap-xx.com/products/prismatic-overview/ [Accessed: November 17, 2024]

[73] Chen S, Rodriguez-Villegas E. Solar energy harvesting node for battery-free physiological monitoring wearable wristband*. In: IECON 2023- 49th Annual Conference of the IEEE Industrial Electronics Society [Internet]. Singapore, Singapore: IEEE; 2023. pp. 1-6. Available from: https://ieeexplore.ieee.org/abstract/document/10312620 [Accessed: February 28, 2024]

[74] Chen ACH, Huang YY, Sharma SK, Hamblin MR. Effects of 810-nm laser on murine bone-marrow-derived dendritic cells. Photomedicine and Laser Surgery. 2011;**29**(6):383-389

[75] Fuada S, Särestöniemi M, Katz M, Soderi S, Hämäläinen M. Experimental study of in-body devices misalignment impact on light-based in-body communications. In: Särestöniemi M, Keikhosrokiani P, Singh D, Harjula E, Tiulpin A, Jansson M, et al., editors. Digital Health and Wireless Solutions. Cham: Springer Nature Switzerland; 2024. pp. 451-466

[76] Yang SY, Sencadas V, You SS, Jia NZX, Srinivasan SS, Huang HW, et al. Powering implantable and ingestible electronics. Advanced Functional Materials. 2021;**31**(44):2009289

[77] Wang K, Karan SK, Sanghadasa M, Wu C, Priya S. Implantable photoelectronic charging (I-PEC) for medical implants. Energy Reviews. 2022;**1**(2):100006

Chapter 4

Perspective Chapter: Designing Urban Sustainability – Leveraging Vehicular Potential Energy for Electric Vehicle Charging and Smart Cities

Tee Hui Teo, Chiang Liang Kok, Zyon Tham, Guangming Ren and Xinlong Zhang

Abstract

This chapter explores innovative advancements in energy systems that address critical challenges in sustainability and efficiency. It examines modern energy storage technologies, highlighting their role in stabilizing renewable energy supplies and enabling applications ranging from grid integration to electric vehicles. The concept of gravitational energy harvesting in multilevel parking structures is presented as a sustainable urban solution, showcasing the conversion of vehicular potential energy into electricity. Power-electronic technologies are discussed for their indispensable role in integrating renewable energy into modern grids, focusing on innovations that enhance efficiency, scalability, and grid stability. Lastly, a hybrid system combining solar energy harvesting with wireless charging demonstrates the potential of renewable energy in creating efficient and resilient sensor networks. Collectively, these advancements underscore the importance of interdisciplinary solutions in achieving a sustainable energy future.

Keywords: energy storage, renewable energy integration, gravitational energy harvesting, power-electronics, solar energy, wireless charging, sustainability

1. Introduction

The imperative of ensuring the sustainability of energy sources cannot be overstated in the current global context. It is crucial to adopt innovative technologies that not only enhance efficiency but also minimize waste and seamlessly integrate renewable energy sources across various sectors. However, several challenges persist in this endeavor, such as the intermittency of renewable energy, grid integration complexities, and the need for efficient energy storage solutions. The current study addresses these pressing issues by delving into four key advancements that are pivotal in reshaping the future of energy systems. Firstly, it explores modern energy storage technologies [1], like batteries and thermal systems, emphasizing their crucial role in stabilizing renewable energy supplies and enabling seamless grid integration. This aspect directly tackles the challenge of intermittency by providing reliable storage solutions for fluctuating renewable energy sources. Secondly, the study introduces gravitational energy harvesting within multilevel parking structures as a novel approach to sustainable urban infrastructure [2]. This innovative concept not only addresses energy generation challenges but also contributes to sustainable development in urban settings. Furthermore, the study discusses the critical role of power-electronic technologies in optimizing renewable energy grid integration [3]. By highlighting innovative solutions that enhance efficiency and tackle integration complexities, the research directly addresses the challenges associated with grid integration of renewable energy sources. Lastly, the study shows a hybrid system that combines solar energy harvesting with wireless charging for sensor networks [4]. This approach not only promotes renewable energy utilization but also enhances network sustainability and resilience, addressing the need for efficient energy solutions in diverse applications.

In essence, this study underscores the importance of sustainable energy practices, outlines existing challenges, and offers innovative solutions to address these issues effectively, paving the way for a more sustainable and efficient energy future.

2. Modern energy storage systems: Characteristics and application in sustainability

Modern energy storage systems are pivotal in enabling sustainable energy solutions, offering key functionalities like stabilizing renewable energy supplies and optimizing grid performance, **Figure 1**. These systems exhibit critical attributes such as energy density, power density, efficiency, and environmental impact, making them suitable for diverse applications [1]. Challenges like cost, scalability, and material constraints persist, necessitating continuous innovation. Various types of energy storage technologies include mechanical systems (e.g., pumped hydro and flywheels),

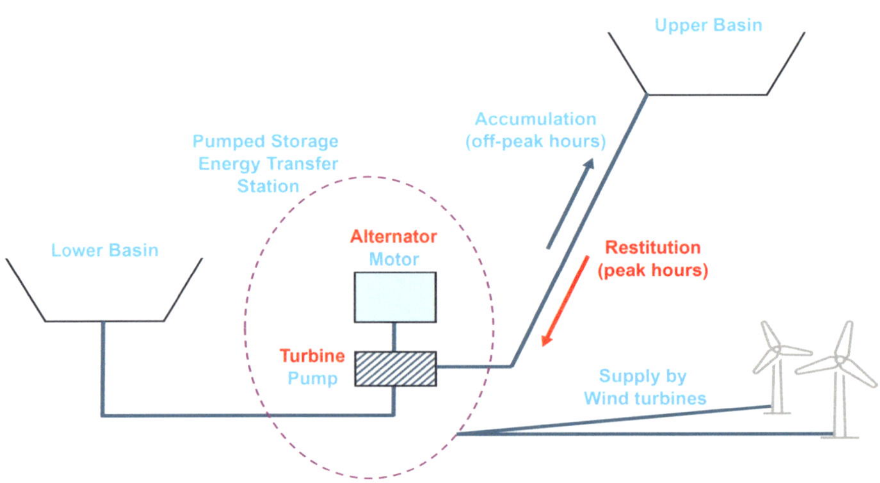

Figure 1.
Sustainable energy storage example [1].

chemical systems (e.g., lithium-ion and flow batteries), and thermal systems (e.g., molten salt and latent heat storage). Each technology has unique strengths tailored to specific use cases, from residential solar energy storage to large-scale grid stabilization and electric vehicles.

By comparing these storage systems, it becomes evident that no single solution can address all energy challenges. Technologies like pumped hydro offer high capacity but face geographic limitations, while lithium-ion batteries provide compact storage with high efficiency but rely on finite resources. Innovations in flow batteries and thermal storage systems highlight scalability and cost-effectiveness for certain applications. Modern energy storage systems play a central role in enabling renewable energy adoption and addressing energy demands across diverse sectors, laying the foundation for a sustainable energy future [1].

3. A sustainable approach to gravitational energy in multilevel parking structure

Gravitational energy harvesting in multilevel parking structures represents an innovative approach to converting potential energy from descending vehicles into usable electricity [2]. This system is illustrated in **Figure 2**. Systems utilizing dynamic droppers, DC generators, and advanced sensors transform this energy into power that can be used for lighting, ventilation, or electric vehicle charging. A proof-of-concept prototype demonstrated the feasibility of this approach, achieving energy outputs that scale significantly for larger parking structures. The mechanism emphasizes design considerations such as speed control, load matching, and modular scalability to maximize energy efficiency and durability. The integration of such systems into urban infrastructure supports sustainable energy production and cost reduction, aligning with the broader goals of energy efficiency and environmental responsibility.

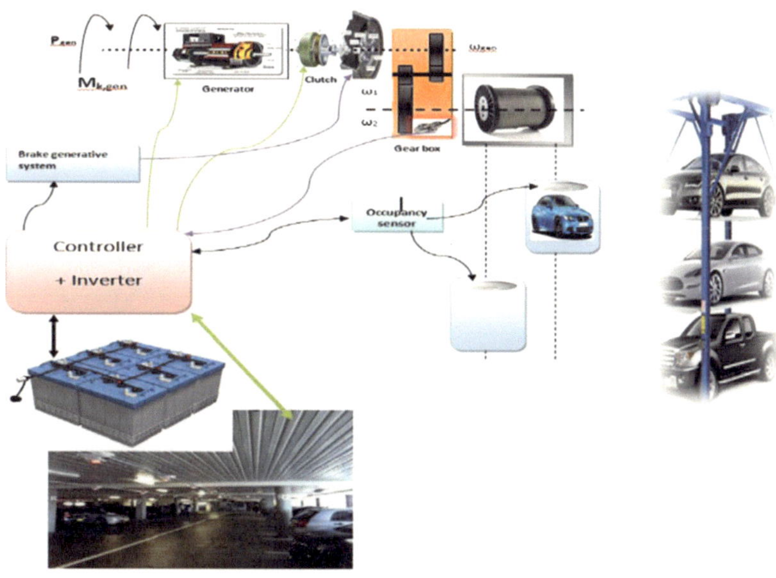

Figure 2.
Gravitational energy harvesting in multilevel parking structures [2].

Challenges like initial investment, maintenance, and retrofitting existing structures remain, but advancements in design and materials continue to address these barriers. By leveraging the significant energy potential of urban vehicle movement, this technology provides a promising pathway for enhancing sustainability and energy independence in densely populated areas.

4. Integration of power-electronic technologies into sustainable energy grid

Power-electronic systems are essential for the seamless integration of renewable energy sources into modern grids. They perform critical functions like energy conversion, stabilizing voltage and frequency, and managing power quality. These systems enable bidirectional power flow, supporting applications such as microgrids and energy storage integration [3]. However, challenges such as intermittency management, scalability, and energy losses remain, necessitating innovative solutions. Technologies like wide-bandgap semiconductors, modular multilevel converters, and smart inverters enhance system efficiency and adaptability, while grid-forming inverters support grids with high renewable penetration. Applications of power electronics [4–8] span solar and wind energy systems, energy storage, microgrids, and hybrid configurations [9–15]. Integration of electric vehicles (EV), as part of the energy grid, is inevitable as it is one of the heavy users of urban energy. To have a good balance between energy usage, traffic, and energy storage, a traffic-aware EV charging management system is proposed in Ref. [9]. The structure of the system is depicted in **Figure 3**.

Advanced control algorithms and materials have made these technologies indispensable for renewable integration, addressing variability and ensuring stable energy delivery. Future trends include AI-driven optimization [6–8], decentralized energy

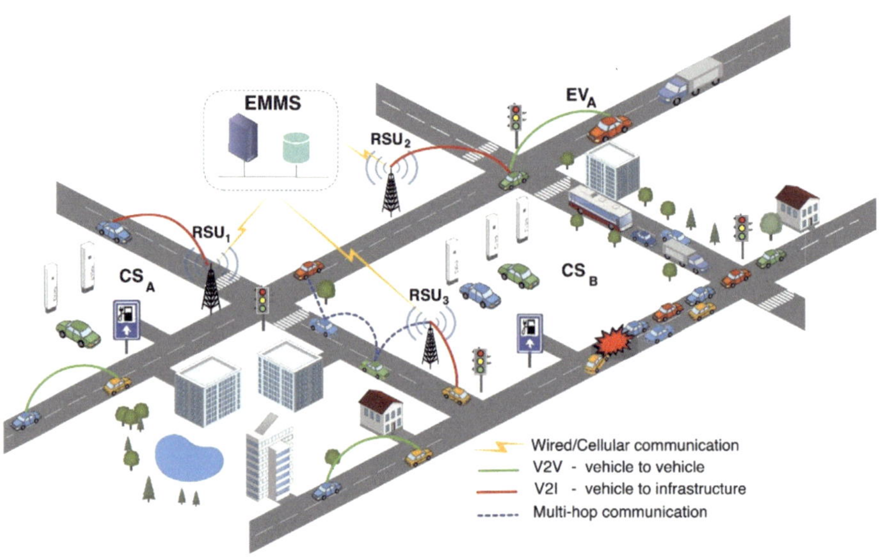

Figure 3
EV charging management system (electro-mobility management server (EMMS)) [9].

systems, and solid-state transformers, further reinforcing the transformative role of power-electronic systems in achieving global sustainability goals [16].

5. A hybrid system approach to solar energy generation and wireless power systems for sensor networks

A hybrid framework combining solar energy harvesting with wireless charging addresses the energy demands of sensor networks as shown in **Figure 4** [4]. Solar-powered cluster heads equipped with photovoltaic panels provide energy for high-demand tasks like data aggregation, reducing dependency on mobile chargers. Wireless-powered nodes, meanwhile, are supported by mobile chargers optimized for efficient routing and energy replenishment. Adaptive strategies, such as dynamic cluster head selection and load balancing, ensure network resilience even during periods of low sunlight.

This hybrid system leverages the strengths of renewable energy and wireless technologies to achieve energy efficiency and sustainability, **Figure 3**. Solar harvesting minimizes environmental impact, while wireless charging provides flexibility and scalability for diverse sensor deployments. Together, these technologies enable robust and continuous network operations, supporting applications in smart cities, agriculture, environmental monitoring, and even healthcare. This hybrid system is potentially to support the energy source for real-time monitoring to enhance reliability, taking inspiration from Refs. [17–19] that utilize ultra-low power sensor node for real-time monitoring. The energy produced by the hybrid system is more than sufficient for the real-time monitoring system, based on estimates where the power consumption is merely 700 μWatt [17–19]. At the same time, for advancement in solid-state passive components such as the planer transformer, spiral inductor [20] also helps in miniaturizing the wireless power. By optimizing energy use and enhancing network reliability, this approach exemplifies the potential of integrated energy systems in advancing sustainable technology.

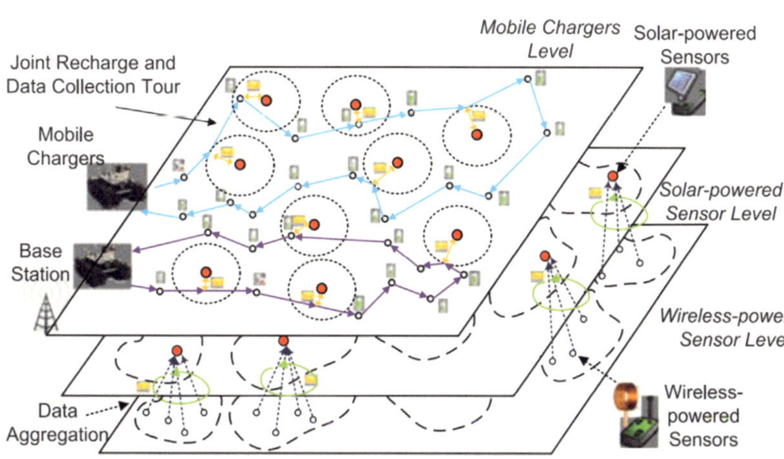

Figure 4.
Hybrid solar power collection system [4].

6. Conclusions

This chapter has explored transformative solutions across various domains of sustainable energy, demonstrating the critical role of innovation in shaping a more resilient and efficient energy future. By examining modern energy storage systems, we highlighted their diverse applications, from stabilizing renewable energy supplies to enabling efficient grid integration and optimizing energy consumption. Gravitational energy harvesting in multilevel parking structures showcased a novel approach to urban sustainability, converting otherwise wasted potential energy into electricity to support infrastructure needs. Power-electronic technologies were identified as pivotal enablers of renewable energy integration, addressing the challenges of variability, scalability, and efficiency while supporting advanced applications like microgrids and hybrid systems. Finally, the hybrid system combining solar energy harvesting with wireless charging underscored the potential of integrating renewable energy with cutting-edge technologies to enhance efficiency and sustainability in sensor networks. Collectively, these discussions illustrate the interconnectedness of technology and sustainability, emphasizing that achieving a reliable and eco-friendly energy future requires multifaceted solutions tailored to diverse applications. The innovations and strategies presented not only address current energy challenges but also lay the foundation for continued advancements, ensuring that energy systems can meet the growing demands of modern society while preserving the environment for future generations.

Acknowledgements

We want to thank the SUTD-ZJU IDEA Visiting Professor Grant (SUTD- ZJU (VP) 202103 and the SUTD-ZJU Thematic Research Grant (SUTD-ZJU (TR) 202204) for supporting this work. This work is supported by DEEPS-AI (Design, Engineering, Education, Psychology, and Science for Artificial Intelligence) initiative.

Author details

Tee Hui Teo[1*], Chiang Liang Kok[2], Zyon Tham[2], Guangming Ren[2] and Xinlong Zhang[2]

1 Singapore University of Technology and Design, Singapore

2 The University of Newcastle, Australia

*Address all correspondence to: tthui@sutd.edu.sg

References

[1] Ibrahim H, Ilinca A, Perron J. Energy storage systems—Characteristics and comparisons. Renewable and Sustainable Energy Reviews. 2008;**12**:1221-1250. DOI: 10.1016/j.rser.2007.01.023

[2] Alkubaisi Y, Hasan WZW, Noor SBM, Kadir MZAA, Sabry AH. Gravitational electric energy of a multilevel parking buildings in commercial and residential sectors proof of concept and prototype measurements. Journal of Computational and Theoretical Nanoscience. 2017;**14**:6030-6037. DOI: 10.1166/jctn.2017.7052

[3] Carrasco JM, Franquelo LG, Bialasiewicz JT, Galvan E, Guisado RCP, Prats M, et al. Power-electronic Systems for the Grid Integration of renewable energy sources: A survey. IEEE Transactions on Industrial Electronics. 2006;**53**:1002-1016. DOI: 10.1109/TIE.2006.878356

[4] Wang C, Li J, Yang Y, Ye F. Combining solar energy harvesting with wireless charging for hybrid wireless sensor networks. IEEE Transactions on Mobile Computing. 2018;**17**:560-576. DOI: 10.1109/TMC.2017.2732979

[5] Khele A, Jiang C, Wang H. Fairness-aware optimization of vehicle-to-vehicle interaction for smart EV charging coordination. IEEE Transactions on Intelligent Transportation Systems. 2023;**24**(4):1234-1245

[6] Schläpfer M, Chew HJ, Yean S, Lee B-S. Using mobility patterns for the planning of vehicle-to-grid infrastructures that support photovoltaics in cities. IEEE Transactions on Sustainable Energy. 2021;**12**(3):1890-1901

[7] Nguyen DM, Kishk MA, Alouini M-S. Modeling and analysis of dynamic charging for EVs: A stochastic geometry approach. IEEE Transactions on Vehicular Technology. 2020;**69**(11):12345-12357

[8] Amilia N, Palinrungi Z, Vanany I, Arief M. Designing an optimized electric vehicle Charging Station infrastructure for urban area: A case study from Indonesia. IEEE Access. 2022;**10**:123456-123467

[9] Bautista PB, Cárdenas LL, Aguiar LU, Igartua MA. A traffic-aware electric vehicle charging management system for smart cities. IEEE Transactions on Vehicular Technology. 2021;**70**(5):1234-1245

[10] Pallarés VF, Cebollada JCG, Martínez AR. Interoperability network model for traffic forecast and full electric vehicles power supply management within the Smart City. IEEE Access. 2021;**9**:12345-12356

[11] Shafie-khah M et al. Optimal behavior of electric vehicle parking lots as demand response aggregation agents. IEEE Transactions on Smart Grid. 2016;7(6):2654-2665

[12] Kok CL, Kusuma IMBP, Koh YY, Tang H, Lim AB. Smart aquaponics: An automated water quality management system for sustainable urban agriculture. Electronics. 2024;**13**:820. DOI: 10.3390/electronics13050820

[13] Kumar AKSB, Kothari DP. Renewable energy for sustainable development in India: Current status, future prospects, challenges, employment, and investment opportunities. IEEE Transactions on Sustainable Energy. 2020;**11**(3):1657-1664

[14] Hannan MA et al. A review of internet of energy based building energy management systems: Issues and recommendations. IEEE Access. 2018;**6**:38997-39014

[15] Liasi SG, Golkar MA. Optimal allocation of electric vehicle charging stations utilizing V2G Technology in Smart Cities. IEEE Transactions on Smart Grid. 2017;**8**(5):2232-2241

[16] Teo TH, Xiang M, Goh E, Hsu H-K. Open-AI driven open-source open-access sustainable ICs design flow. In: 2024 IEEE 17th International Symposium on Embedded Multicore/Many-Core Systems-on-Chip (MCSoC). Kuala Lumpur, Malaysia: IEEE; 2024. pp. 361-365. DOI: 10.1109/MCSoC64144.2024.00066

[17] Teo TH, Lim GK, David DS, Tan KH, Gopalakrishnan PK, Singh R. Ultra low-power sensor node for wireless health monitoring system. In: 2007 IEEE International Symposium on Circuits and Systems (ISCAS). New Orleans, LA, USA: IEEE; 2007. pp. 2363-2366. DOI: 10.1109/ISCAS.2007.377934

[18] Teo TH et al. A 700-μW single-chip IC for wireless continuous-time health monitoring in 0.18-μm CMOS. In: Proc. 2009 IEEE Asian Solid-State Circuits conference. Taipei, Taiwan: IEEE; 2009. pp. 361-364. DOI: 10.1109/ASSCC.2009.5357170

[19] Teo TH et al. A 700-μW wireless sensor node SoC for continuous real-time health monitoring. IEEE Journal of Solid-State Circuits. 2010;**45**(11):2292-2299. DOI: 10.1109/JSSC.2010.2064030

[20] Teo TH, Choi Y-B, Liao H, Xiong Y-Z, Jeffrey Shiang F. Characterization of symmetrical spiral inductor in 0.35 μm CMOS technology for RF application. Solid-State Electronics. 2004;**48**(9):1643-1650. DOI: 10.1016/j.sse.2004.04.004

Chapter 5

Perspective Chapter: Advancements in Wireless Charging – A Comparative Study of Emerging Technologies

Tee Hui Teo, Chiang Liang Kok and Hongyi Goh

Abstract

This chapter delves into the profound impact of solar energy on the progression of wireless charging technology across various applications. It explores a wide array of innovations, including portable solar-powered charging devices that offer convenience and sustainability, multiload charging systems that cater to diverse needs, hybrid charging designs tailored for wireless sensor nodes, and cutting-edge BIPV-powered drone charging systems. These advancements underscore the transformative potential of solar energy in bolstering the efficiency, autonomy, and environmental sustainability of wireless charging technologies. This chapter highlights the synergistic relationship between solar energy and wireless charging technology, showcasing how solar power significantly enhances the efficiency, autonomy, and sustainability of various wireless charging applications. The advancements discussed, from portable chargers to drone charging stations, demonstrate the potential for a cleaner, more efficient, and readily accessible future powered by the sun.

Keywords: wireless charging technology, solar-powered, sustainability, smart charging infrastructure, energy harvesting, power efficiency

1. Introduction

The goal for this chapter is to provide a comparative study of the emerging technology in wireless charging and propose an integrated solution with solar energy. Wireless charging technology has evolved significantly, reshaping the landscape of energy solutions for modern devices and systems. These technologies promise increased convenience, efficiency, and sustainability, addressing the growing demand for seamless energy transfer. The primary appeal of wireless charging is demonstrated through its technique to eliminate the need for physical connectors, offering convenience and flexibility. Wireless charging systems have also evolved to integrate renewable energy sources, such as solar power, making it an ideal solution for off-grid or remote locations. This chapter will explore the current innovations in wireless charging technologies, starting from portable solar-powered wireless chargers. These devices are especially beneficial in regions with limited access to reliable electricity, ensuring users get to power their devices sustainably. The battery life of power devices can now be extended through advanced algorithms such as maximum power point tracking (MPPT), which enhances the performance of these chargers by optimizing energy harvesting from solar panels [1]. Another area of development is multiload constant current charging systems, as this technology can charge multiple devices simultaneously with high efficiency, even under fluctuating power conditions. In environments where it requires several devices to be charged at the same time, this ability is crucial for managing varying energy demands [2].

For larger-scale distributed environments, the chapter will look into solar-powered wireless sensor nodes, as these sensor nodes can operate autonomously, harvesting solar energy to power itself sustainably [3]. Building Integrated Photovoltaic (BIPV) systems for drones will also be essential for future autonomous urban drone transportation and delivery by combining solar energy harvesting with wireless charging to power drone fleets in urban environments [4]. Not excluding the healthcare sector, with wireless charging technology providing seamless, efficient power solutions for active implantable and wearable medical devices, enhancing both functionality and patient convenience [5, 6]. Finally, the growing adaptation of EVs has accelerated the development of wireless charging systems, offering different solutions to improve energy efficiency, sustainability, and user accessibility in modern transportation [7].

2. Foundation of wireless charging technologies

Wireless charging is the transfer of energy from a power source to a device without the use of physical connectors. Over the years, the principles of electromagnetic induction and resonant coupling technologies that form the foundation of modern wireless energy systems have evolved, offering more efficient and versatile solutions for consumers and industrial applications. Widely used methods:

2.1 Inductive coupling

Relies on electromagnetic induction to transfer energy between two coils, typically used for short-range or lower-power applications such as smartphones.

2.2 Resonant coupling

Allows energy transfer over greater distances and is less sensitive to alignment, ideal for large-scale applications such as charging multiple devices or powering drones and sensor networks.

With wireless charging technologies evolving rapidly, reducing the dependence on the traditional energy grid has been the focus by integrating renewable energy sources such as solar power.

3. Innovations in portable wireless charging systems

Portable solar-powered wireless chargers mark a significant milestone in wireless energy solutions. By using photovoltaic (PV) panels to capture sunlight, it will be converted to electrical energy before transferring wirelessly to compatible devices. To ensure that the solar panel operates at peak efficiency regardless of the light conditions, the use of MPPT algorithms can help optimize the energy harvesting process. Through this, remote or off-grid locations where conventional power sources are limited can now rely on portable solar chargers, as they offer sustainable and eco-friendly solutions for powering small devices [2, 3]. Furthermore, multiload constant current charging systems have emerged as an essential innovation for powering multiple devices simultaneously. In industrial settings or emergency response operations where multiple devices are required to be charged at once, these systems ensure each device receives a stable and reliable power supply. Maintaining high efficiency with the ability to charge multiple devices under varying load conditions is a significant improvement over traditional charging methods, as it avoids issues like overcharging, heat generation, or power fluctuations [2].

4. Wireless sensor nodes with hybrid charging and storage

Wireless Sensor Networks (WSNs) have become integral to modern infrastructure, supplying vital information in applications from smart grid management to environmental monitoring. However, the limited power available to these devices has frequently impeded the implementation of large-scale WSNs. Solar-powered sensor nodes have emerged as a promising solution to tackle this problem. These nodes may function independently without requiring external power sources since they integrate hybrid charging and energy storage systems with solar energy harvesting [3], as depicted in **Figure 1**.

A key innovation in this field is the use of dual energy storage systems that combine supercapacitors and lithium batteries. Lithium batteries store energy for extended periods of time, guaranteeing continued operation even on cloudy days or

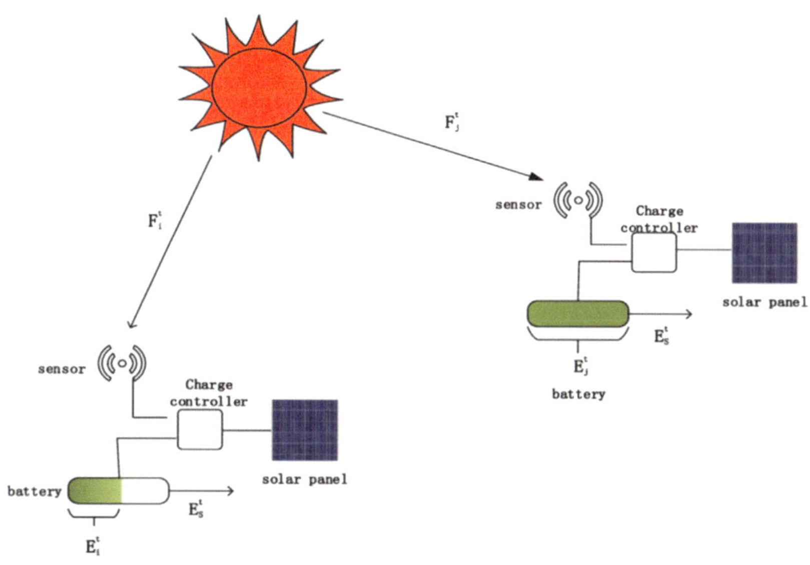

Figure 1.
Solar sensor node structure (radiated energy from solar: F_i^t, F_j^t; supercapacitor energy: E_i^t, E_j^t; lithium battery energy: E_s^t) [3].

at night. Supercapacitors, however, can charge and discharge rapidly, making them perfect for handling brief energy bursts from intense sunlight. The ability to switch between these two storage modes based on available sunlight ensures that the sensor node can maintain a reliable power supply [3]. Another essential part of this system is the power management circuit. It features an undervoltage lockout (UVLO) to prevent the system from running out of power and an MPPT to maximize the energy collected from solar panels. The storage system is always charged and ready to use thanks to the hybrid charging system's ability to manage varying current levels based on the intensity of sunlight [3].

5. BPIV-powered drone charging systems

The integration of renewable energy systems such as BIPV is becoming increasingly important as urban environments become more technologically advanced. BPIV systems offer a sustainable means of powering a variety of devices, including drones, by seamlessly combining energy generation with building infrastructure. For urban drone delivery services, which require effective, self-sufficient charging options, these systems are necessary. BIPV systems provide a flexible way to charge drone fleets by incorporating PV panels into building structures like walls and rooftops, ensuring that the energy required for operation is easily accessible and renewable (**Figure 2**) [4]. According to a study on wireless drone charging systems driven by BIPV, specific panel layouts and orientations can significantly improve charging efficiency. For instance, it was demonstrated that a drone battery could be charged in 31.29 minutes as compared to 36.9 minutes with vertical panels. In comparison to white-reflecting surfaces, the energy yield was further enhanced by 10.09% when black reflective surfaces were used. These results highlighted the importance of optimizing panel orientation to maximize energy generation [4]. Effective power transfer is made possible even across a vertical displacement of 0–10 mm by the incorporation of a 120 W wireless charging technology with an efficiency rate of 95.7%. This innovation is essential for the future of drone-based delivery systems as

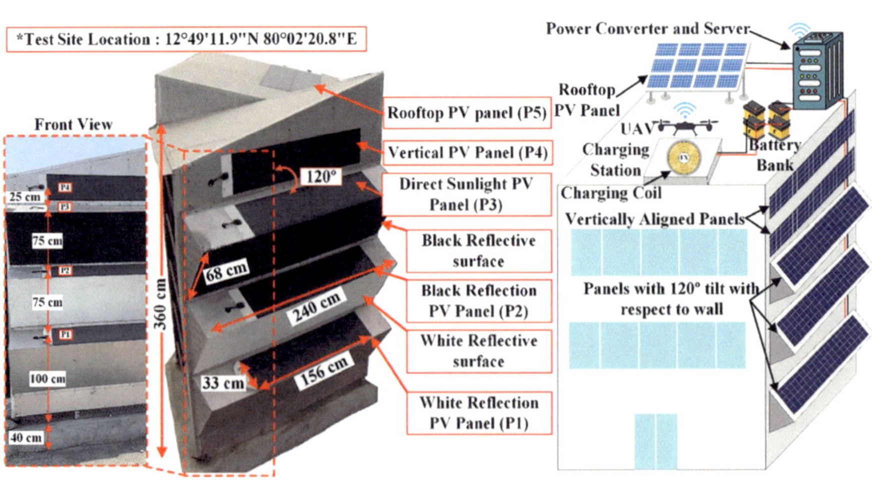

Figure 2.
BIPV structure (left), application of wireless drone charging (right) [4].

it enables drones to charge automatically and smoothly in urban settings. To meet the increasing demand for urban drone services, this article [4] investigates how BIPV-powered drone charging systems can offer a viable alternative for the development of autonomous vehicles in the future.

6. Wireless charging technology for medical devices

Wireless charging technology has become a cornerstone for improving medical devices, particularly in active implantable medical devices (AIMDs) and wearable devices [5, 8]. Recent developments open the door to improved patient convenience and device functionality by addressing the crucial requirements of safety, power efficiency, small size, and smooth communication. Wireless charging solutions for AIMDs, such as cardiac pacemakers and neurostimulators, have undergone substantial advancements. These devices require reliable, compact, and biocompatible power solutions due to their operation in highly sensitive physiological environments. Some of the current advancements include bidirectional communication modules, which enable the wireless charging system to exchange critical data with external monitoring systems, including device status and bio-signals, in addition to supplying power [9]. These systems make use of inductive coupling techniques that are designed for patient safety and power transmission efficiency. Safety features, such as thermal regulation and overcharge prevention, are integrated to ensure stable performance without compromising the device's longevity. A layer of operational reliability is added by the bidirectional communication capability, which is accomplished using frequency-shift keying and other methods, making these devices more intelligent and responsive [6].

Wearable medical devices, like hearing aids and glucose monitors, emphasize convenience, tiny form factors, and ultra-low power. These requirements have been successfully met by the creation of a wireless power charger integrated circuit (IC) using BCDMOS technology [7], as depicted in **Figure 3**. This setup uses cross-coupled nMOS transistors and Schottky diodes to create a passive full-wave rectifier [10]. This design minimizes voltage drops, enhances power efficiency, and prevents reverse leakage currents, making it ideal for the compact and low-power demands of wearable devices. Bulkier switching circuits are replaced with a linear charging circuit to further fit the tiny form factor. Additionally, in-band communication circuits

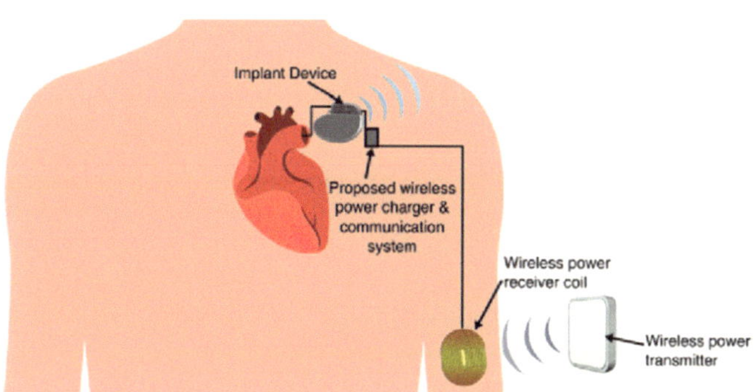

Figure 3.
Implantable medical device with wireless charging [7].

enable real-time updates on battery status and connection integrity through the same wireless power channel, simplifying the overall design and reducing component complexity. Nickel-Metal Hydride (NiMH) batteries are considered over lithium-ion alternatives to reduce the risk of overcharging or thermal incidents, which are critical for patient safety.

Wireless charging on AIMDs prioritizes biocompatibility, minimal invasiveness, and strong safety measures. Wireless charging on wearable medical devices, however, places an emphasis on portability, user-friendliness, and effective communication integration [11]. Improvements in rectifier design, heat management, and battery safety have benefited both fields, but AIMDs are the first to include advanced bidirectional connectivity for real-time monitoring. By improving power transfer efficiency, integrating real-time communication, and ensuring patient safety, these advancements mark a significant step toward more convenient and effective healthcare solutions.

7. Wireless charging for electric vehicles (EVs)

Electric Vehicles (EVs) have been dominating the global market for a few years, and the advancements of wireless charging technology for them are gaining even more attention as a promising solution to address energy efficiency and carbon emission challenges. To optimize the power density and minimize interlayer voltage stress, a dual-layer receiver for EV wireless charging systems was proposed (**Figure 4**) [12]. By using a dual-layer structure, this innovative receiver design addresses the prevalent problem of low power density in conventional single-layer systems. The architecture is designed to ensure efficient power transfer while addressing safety concerns, such as interlayer voltage stress, through the strategic use of insulation and optimized design parameters. The experimental findings showed that the suggested method outperformed traditional designs in terms of power density and stress, representing

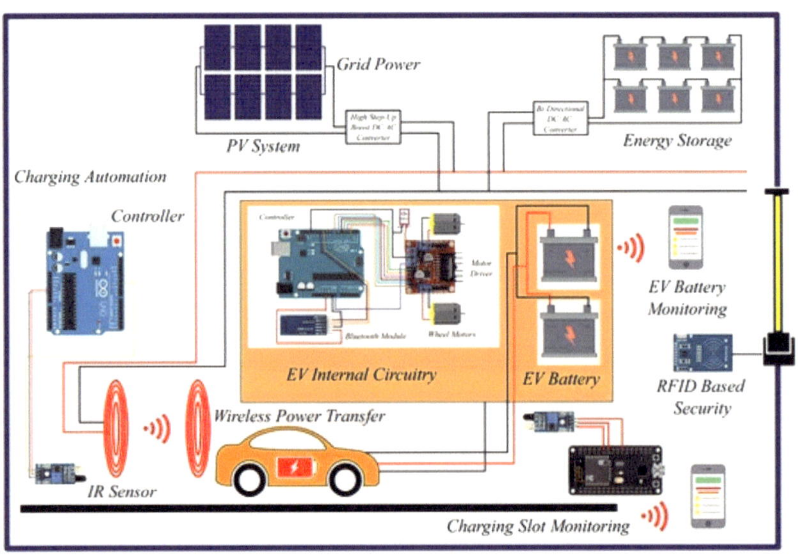

Figure 4.
Integration of renewable energy and IoT for advanced wireless power transfer in EVs [11].

a major advancement in EV wireless charging technology [12]. Additionally, the system's adaptability to different EV models and its potential scalability make it a strong contender for broad adoption in future EV wireless charging infrastructure.

Another advancement in wireless charging systems of EVs is utilizing deep reinforcement learning (DRL) methods to incorporate renewable energy sources, such as solar and wind. The importance of incorporating renewable energy resources to enhance the sustainability of EV charging while mitigating environmental impacts is highlighted. To improve the charging process while accounting for the uncertainty in energy generation and consumption patterns, it presents a constrained deep double Q-network (CDDQN) framework. Using a Markov Decision Process (MDP) to formulate the charging problem, the study successfully addressed the intermittent renewable energy supply challenge. The proposed DRL-based solution enhanced charging efficiency by reducing Q-value estimation errors and ensuring reliable charging coordination for multiple EVs. The system's robustness in managing fluctuating energy supply and demand was shown through simulations, paving the path for more sustainable wireless charging solutions [13]. The integration of IoT into wireless charging stations further transforms user experiences. IoT-enabled stations can provide real-time updates on grid health and charging progress while allowing remote monitoring and scheduling through mobile applications. This connectivity fosters user predictability and control, which are crucial for mass adoption. The RFID systems bolster security by restricting access to authorized vehicles, ensuring safe and trusted operations [14]. With all these advancements, it highlights the pivotal role of IoT, RERs, and advanced algorithms in shaping the future of wireless EV charging, addressing associated energy demands with innovative, sustainable solutions.

8. Future direction of wireless charging technologies

The field of wireless power transfer (WPT) has advanced significantly, particularly in the context of dynamic charging for EVs, from the small-scale handheld wireless portable charger to larger-scale dynamic charging. Wireless power distribution has historically been used for stationary equipment, but as demand for EVs and other mobile gadgets rises, innovative solutions have emerged to deliver energy efficiently to moving systems. Wireless dynamic charging for electric busses is an example of research presented by researchers in Auckland on an innovative approach for delivering wireless power to EVs while in motion [15]. It describes the development of a wireless charging system that uses a dynamic charging lane that enables the car to receive power without the need to stop. The key element of this system is the implementation of a track embedded with primary coils, which generates an electromagnetic field, and secondary coils that absorb energy under the vehicle. This strategy tackles several important issues, such as the requirement for a steady power source while moving and the optimization of energy transfer when the car moves over the charging lane. It also demonstrates that the system's efficiency is still comparatively high even at the speed of 50 km/h, opening possibilities for future EV infrastructures that allow busses to continue operating while charging at the same time. Some of the challenges, such as optimal alignment between coils and minimizing power loss due to vehicle speed and position, are areas that need to be investigated when creating the system [15]. Further work is required to optimize the system for different types of vehicles and real-world conditions, such as the design of a dynamic wireless charging power track, as depicted in **Figure 5**.

(a) DWC-EV power track under construction (b) DWC-EV in operation

Figure 5.
Dynamic wireless charging power track [16].

Another study shows that by using resonant inductors that exchange energy between them and coupling them with an amplifier, the system can automatically adjust its frequency for maximum power transfer [17]. This strategy fundamentally shifts the way wireless energy is transferred by allowing the system to remain efficient even with the movement of both the receiver and transmitter, regardless of the distance between the coils, up to 70 cm. This is a novel approach to overcoming the limitations of static wireless power transmission by applying a concept from quantum mechanics known as parity-time symmetry. Static coils and precisely calibrated electronic circuits, which are usually required for effective energy transfer, are no longer needed after modification of the dynamic operating frequency. This advancement could revolutionize the way energy is delivered to moving devices like EVs, enabling a more flexible and durable charging infrastructure.

All this research highlights how renewable energy [18], especially solar electricity, can improve power efficiency [19, 20], sustainability [21], and the autonomy of wireless charging technology [22–24]. Through its tailored solutions for medical [5, 8] and automotive applications, it represents a transformative step toward a safer, more efficient, and more sustainable energy transfer system.

9. Conclusions

In conclusion, these studies highlight the growing significance of renewable energy in the advancement of wireless charging technology. Each article focuses on how solar energy can be integrated to create sufficient, sustainable energy solutions. Solar-powered wireless chargers can offer eco-friendly alternatives, especially in areas with restricted access to electricity. Multiload charging system expands the concept and guarantees effective power distribution to several devices, which is critical for wireless network scalability. By using a hybrid design that combines supercapacitors with lithium batteries for continuous operations, it demonstrates the importance of energy storage and power management for autonomous systems even in varying environmental conditions. Solar energy integration into urban infrastructure can support autonomous, sustainable drone operations with the potential of BIPV-powered wireless drone charging systems. Wireless charging on medical devices focuses on enhancing patient safety, biocompatibility, and efficiency, with implantable devices prioritizing bidirectional communication and wearable devices emphasizing compactness and user convenience. Dual-layer receivers and DRL-based frameworks are two examples of wireless EV charging developments that maximize power density, improve sustainability through the integration of renewable energy, and employ RFID and IoT to improve user experience and security. By developing further, it offers promising solutions for reducing reliance on traditional energy sources and advancing the next generation of wireless charging.

Acknowledgements

We want to thank the SUTD-ZJU IDEA Visiting Professor Grant (SUTD-ZJU (VP) 202103) and the SUTD-ZJU Thematic Research Grant (SUTD-ZJU (TR) 202204) for supporting this work. This work is supported by DEEPS-AI (Design, Engineering, Education, Psychology, and Science for Artificial Intelligence) initiative.

Conflict of interest

"The authors declare no conflict of interest."

Author details

Tee Hui Teo[1*], Chiang Liang Kok[2] and Hongyi Goh[2]

1 Singapore University of Technology and Design, Singapore

2 The University of Newcastle, Australia

*Address all correspondence to: tthui@sutd.edu.sg

References

[1] Kok CL, Fu X, Koh YY, Teo TH. A novel portable solar powered wireless charging device. Electronics [Internet]. 2024;**13**(2):403. DOI: 10.3390/electronics13020403

[2] Zhang Z, Xiaojuan Z, Yulei X, Guangjun X, Xin C. Multi-load constant current charging technology for wireless charging systems. International Journal of Electronics [Internet]. 2020;**107**(8):1254-1271. Available from:. DOI: 10.1080/00207217.2020.1726487

[3] Wang L, Li D, Ma Y, Zhao L, Chen Y, Zhang J, et al. Hybrid charging and storage Design in Sustainable Solar-powered Wireless Sensor Node. Sensors and Materials [Internet]. 2022;**34**(7):2771. DOI: 10.18494/sam3740

[4] Chittoor PK, Bharatiraja C. Building integrated photovoltaic powered wireless drone charging system. Solar Energy [Internet]. 2023;**252**:163-175. Available from:. DOI: 10.1016/j.solener.2023.01.056

[5] Teo TH et al. A 700-μW wireless sensor node SoC for continuous real-time health monitoring. IEEE Journal of Solid-State Circuits. 2010;**45**(11):2292-2299

[6] Çetin S, DemiRci Y E, Büyükgümüş O. Performance evaluation of a wireless charging converter for implantable active medical device. Mugla Journal of Science and Technology [Internet]. 2020;**6**(2):11-17. DOI: 10.22531/muglajsci.742801

[7] Asan NB. FAT-IBC: A New Paradigm for Intra-Body Communication. Uppsala: Acta Universitatis Upsaliensis; 2019

[8] Teo TH et al. A 700-μW single-chip IC for wireless continuous-time health monitoring in 0.18-μm CMOS. In: 2009 IEEE Asian Solid-State Circuits Conference. Taipei, Taiwan: IEEE; 2009. pp. 361-364. DOI: 10.1109/ASSCC.2009.5357170

[9] Ju X, Liu W, Zhang C, Liu A, Wang T, Xiong NN, et al. An energy conserving and transmission radius adaptive scheme to optimize performance of energy harvesting sensor networks. Sensors [Internet]. 2018;**18**(9):2885. Available from:. DOI: 10.3390/s18092885

[10] Moon Y-J, Kim S, Choi Y, Lee Y, Park J, et al. Wireless power charger for wearable medical devices with in-band communication. International Journal of Circuit Theory and Applications [Internet]. 2015;**44**(8):1483-1493. DOI: 10.1002/cta.2173

[11] Zhou Y, Liu C, Huang Y. Wireless power transfer for implanted medical application: A review. Energies. 2020;**13**:2837. DOI: 10.3390/en13112837

[12] Liu C, Gao X, Cui S. A dual-layer receiver with low interlayer voltage stress and high-power density for EVs wireless charging system. IEEE Transactions on Industrial Electronics. 2025;**72**(4)3551-3561. DOI: 10.1109/tie.2024.3451140

[13] Balaji S, Anuradha T. Adaptation of renewable energy resources to wireless EV charging using deep learning. Electric Power Components and Systems. 27 Oct 2023:1-15. DOI: 10.1080/15325008.2023.2269928

[14] Iqbal S, Alshammari NF, Shouran M, Massoud J. Smart and sustainable wireless electric vehicle charging strategy with renewable energy and internet of things integration. Sustainability [Internet]. 2024;**16**(6):2487. Available from:. DOI: 10.3390/su16062487

[15] Du B, Ajith VS, Mingyue SS, Prakash R, Le W, Basil S, et al. Optimal deployment of dynamic wireless charging infrastructure for battery electric buses – a case study in Auckland, New Zealand. Journal of the Royal Society of New Zealand. 2024:1-20. DOI: 10.1080/03036758.2024.2353737

[16] Jang Y, Jeong S, Lee M. Initial energy logistics cost analysis for stationary, quasi-dynamic, and dynamic wireless charging public transportation systems. Energies [Internet]. 2016;**9**(7):483. Available from:. DOI: 10.3390/en9070483

[17] Lerosey G. Wireless power on the move. Nature [Internet]. 2017;**546**(7658):354-355. DOI: 10.1038/546354a

[18] Don Chua WF, Lim CL, Koh YY, Kok CL. A novel IoT photovoltaic-powered water irrigation control and monitoring system for Sustainable City farming. Electronics. 2024;**13**:676. DOI: 10.3390/electronics13040676

[19] He Y, Meng F, Kok CL, Ho CK. A K-to-Ka band ultra-wideband power amplifier in 180-Nm SiGe technology. In: 2024 IEEE International Conference on IC Design and Technology (ICICDT). Singapore, Singapore: IEEE; 2024. pp. 1-4. DOI: 10.1109/ICICDT63592.2024.10717834

[20] Li W, Meng F, Kok CL, Ho CK. A 1MHz, 93.7% peak efficiency, heterogenous GaN/BCD 48-to-3.3V~5V DC-DC buck converter. In: 2024 IEEE International Conference on IC Design and Technology (ICICDT). Singapore, Singapore: IEEE; 2024. pp. 1-4. DOI: 10.1109/ICICDT63592.2024.10717805

[21] Kok CL, Kusuma IMBP, Koh YY, Tang H, Lim AB. Smart aquaponics: An automated water quality management system for sustainable urban agriculture. Electronics. 2024;**13**:820. DOI: 10.3390/electronics13050820

[22] Zhang Y, Zhao Z, Yuan L, Li W. Dynamic wireless charging for electric vehicles with autonomous frequency control. IEEE Transactions on Industrial Electronics. 2023;**70**(1):123-134. DOI: 10.1109/TIE.2022.3145678

[23] Kumar A, Singh B, Jain C. Wireless charging of an autonomous drone. IEEE Transactions on Power Electronics. 2020;**35**(12):12345-12354. DOI: 10.1109/TPEL.2020.2991234

[24] Wang H, Liu X, Chen M. Novel autonomous self-aligning wireless power transfer for electric vehicles. IEEE Transactions on Transportation Electrification. 2020;**6**(3):987-997. DOI: 10.1109/TTE.2020.2995678

Chapter 6

Perspective Chapter: Market Dynamics of Pakistan's Energy Transition

Mahmood Ahmad, Haneea Isaad and Furqan Ali Akhtar

Abstract

The global energy transition is accelerating, marked by COP28's historic commitment to triple renewable energy capacity and double energy efficiency by 2030—critical steps toward the 1.5°C climate goal. In 2022–23, renewables accounted for 83% of new power capacity additions, driven by falling costs, economies of scale, and supportive policies. However, questions persist: Are these efforts sufficient to meet climate targets and ensure sustainable development? This chapter examines the diverging renewable energy trajectories of Pakistan and India within this global context. While the Global North sees private-sector-led transitions, China and some countries in the Global South rely on strong public-sector leadership. Pakistan, though gradually shifting toward renewables like solar and wind, remains heavily dependent on imported fossil fuels. Its energy policies face significant implementation challenges due to bureaucratic inertia, financial constraints, and political instability. High energy costs, circular debt, and weak infrastructure further hinder progress, undermining economic competitiveness. In contrast, India has made notable progress, particularly in scaling solar and wind energy, with policy targets and investments surpassing Pakistan's. This chapter explores how Pakistan can align its energy policies to catalyze transformation—by reducing fossil fuel dependence, improving regulatory frameworks, and attracting targeted investments. It also discusses the need for institutional realignment to support a renewable-dominated energy future, and the broader socio-economic and environmental outcomes of a successful transition.

Keywords: renewable energy, energy prices, externalities, policy failures, circular debt

1. Introduction

1.1 Global review

In the northern hemisphere, the high-income countries—such as the United States, Germany, Japan, Canada, and Australia—are renowned for their advanced energy infrastructures. For decades, these nations have enjoyed reliable, affordable, and stable energy access, which has fueled their economic growth and supported high living standard. Their energy portfolios are diverse, comprising renewables like solar and wind, alongside fossil fuels and nuclear power. This diversification has allowed them to maintain energy security even in the face of fluctuating global markets. However, energy markets today carry an uncertain future. The reliable and affordable energy access that high-income countries once took for granted is no longer as assured. The ongoing crises in Ukraine and the Middle East have injected volatility into global energy markets, pushing governments to rethink their energy strategies [1, 2]. While the long-term focus remains on sustainability and decarbonization, the short-term realities of energy security and economic stability have prompted a renewed reliance on fossil fuels in many countries. High-income nations are facing a delicate balancing act: managing the immediate need for stable energy supplies while continuing to push for a cleaner, more resilient energy future.

Emerging economies like China, India, Brazil, and South Africa are experiencing rapid industrialization, which drives their growing energy demands [3]. These nations are more vulnerable to energy price fluctuations and supply disruptions but are also increasingly adopting renewable energy technologies to meet their energy needs. Both China and India are increasingly investing in renewable energy, grid modernization, and energy storage solutions to better integrate renewable energy into their systems and ensure reliability [4, 5]. While still reliant on fossil fuels, they are embracing renewables to reduce vulnerability to global energy market disruptions and make progress on climate commitments. Pakistan though having abundant renewable energy resources faltered on a number of policies resulting in the energy sector turning out to be a heavy fiscal drain on the economy [6].

This chapter shows how Pakistan's energy markets are a classic case of policy, institutional, and market failures. Across the border, India is quickly emerging as a success story in energy transition while Pakistan is still struggling with a multitude of technical and economic challenges which must be overcome before the country embarks on its clean energy pathway.

1.2 Pakistan's energy mix

Pakistan's historical over-reliance on abundant gas and hydroelectric power (30% and 60%, respectively) delayed crucial infrastructure investments. Today, Pakistan's energy mix (as of July 2024) comprises hydel (25.35%), thermal (45%), nuclear (18.19%), and renewable (3.26%), and is diverse, but fossil fuels—such as coal, oil, and natural gas—hold a major share (70%). Further, as shown in **Table 1**, the energy mix has shifted from hydropower as the lead in the 1960s to oil and coal, affecting

Year	Hydropower (%)	Natural gas (%)	Oil (%)	Coal (%)	Nuclear (%)	Renewables (%)	Others (%)
2024	30	40	20	10	6	4	<1
2004	29	50	16	4	1	0	<1
1984	35	45	16	2	2	0	<1
1964	60	30	8	2	0	0	<1

Source: Ministry of Energy, 2023.

Table 1.
How Pakistan's energy mix has changed.

fiscal and environmental policy space. Pakistan's domestic oil and gas resources are limited, while coal mining is still being scaled up, so the country is heavily dependent upon imported fossil fuels (primarily furnace oil and liquefied natural gas (LNG)). This reliance on imported energy exposes the economy to external shocks like global oil and gas price fluctuations, leading to significant costs for the government and consumers. This dependency diverts critical foreign reserves to pay for fuel imports, increasing the country's fiscal deficit. In FY 2022–2023, the energy import bill reached $23 billion, accounting for 30% of the country's total imports [7].

2. Pakistan energy sector—A deeper dive

Pakistan intends to set a cumulative ambitious conditional target of overall 50% reduction of its projected emissions by 2030, with 15% from the country's own resources and 35% subject to provision of international grant finance that would require USD 101 billion just for energy transition [8]. To reach the target, Pakistan aims to shift 60% to renewable energy, and 30% to electric vehicles by 2030 and completely ban imported coal. Furthermore, Pakistan aims to expand nature-based solutions (NbS) by implementation of Ten Billion Tree Tsunami Project (TBTTP) and Recharge Pakistan (RP) [9].

Pakistan has been trying to advance on its clean energy goals through the introduction of consumer-friendly policies. In 2015, the National Electric Regulatory Authority (NEPRA) introduced the Distributed Generation and Net Metering Regulations, establishing a framework for integrating distributed renewable energy. Since then, rooftop solarization has surged, with around 3.2 gigawatts (GW) of net-metered rooftop solar photovoltaic (PV) systems connected to the grid by June 2024. However, the government is now considering revisions to the net metering policy, including lowering buyback rates and transitioning to net billing. While this shift could extend payback periods for consumers with higher self-consumption, it may also encourage the installation of larger solar systems. Pakistan's smart grid initiatives include the Pakistan Electric Power Company (PEPCO) launching a Smart Grid Project in 2018 and several distribution companies adopting advanced metering infrastructure. Although some of the distribution companies are piloting smart grid technologies [10], widespread implementation of digitization at a national level is still lacking. In fact, high transmission and distribution (T&D) losses, system leakages, power theft, and under-investment in transmission and distribution infrastructure has led to persisting grid challenges with consumers suffering load-shedding on a regular basis despite the grid being in surplus. The distribution companies also carry out load shedding due to economic or commercial reasons [11].

The Ministry of Energy has also led an Energy Efficiency and Conservation program since 2019. Despite these efforts, challenges such as interoperability, scalability, security, regulatory frameworks, and lack of public awareness persist. Lately, due to an increased adoption of distributed solar technology across the country, stagnant industrial growth, and high electricity tariffs, electricity demand from the grid has been shrinking over the past 2 years, dropping by 10% in 2023 and 17% in 2024 [12].

Experts suggest optimizing buyback rates for net-metering, promoting hybrid systems with battery storage, and investing in performance monitoring to overcome these hurdles [13]. In addition, the government is actively trying to correct for its over-reliance on imported fuels which is responsible for the extremely high cost of electricity in the country. Pakistan's power tariffs are one of the highest in the region, surpassing India and Bangladesh by a wide margin and going up to 22 US cents/KWh (after taxation) for some consumer categories. In October 2024, the government of Pakistan terminated power purchase agreements (PPAs) with 5 thermal Independent Power Producers (IPPs) to reduce the burden of fixed capacity payments and reduce electricity prices [14].

2.1 Circular debt: A persistent constraint

Pakistan's power sector is crippled by mounting circular debt, which stood at PKR 2.6 trillion by April 2024 [15, 16], severely affecting liquidity and discouraging investment [8]. Circular debt occurs when electricity distribution companies (DISCOs) fail to collect payments from consumers, leading to cash flow issues. These distribution companies are often unable to pay power generation companies, which, in turn, cannot pay for fuel or maintenance. This chain reaction cripples the sector's ability to invest in upgrading infrastructure and ensuring efficient operations.

Figure 1 below highlights how the value chain of circular debt progresses while hampering the creation of healthy backflow of cash.

There are number of macro variables that make the equation worst; the lead factor being the linking of energy process to the US dollar through indexation throughout the energy value chain. Hence, the PKR to USD exchange rate plays a critical role in determining imported fuel costs which the sector heavily relies on. Between 2015 and 2023, the Pakistani rupee (PKR) depreciated substantially, from PKR 101 to PKR 300 per USD, more than doubling local currency oil import costs [17]. This depreciation affects inflation, trade balance, and fiscal stability resulting in high import costs, which are ultimately passed down to the consumers through tariff adjustments.

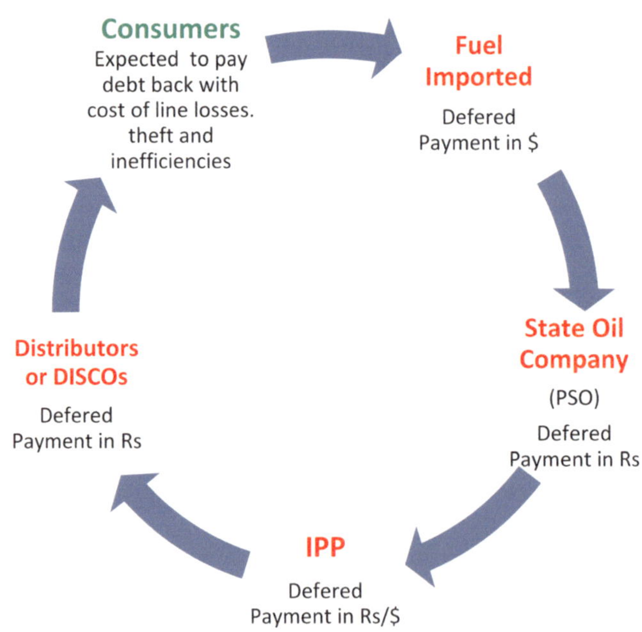

Figure 1.
How circular debt is generated.

3. Analytical framework

The problems of energy shortage, loadshedding, circular debt, aging infrastructure, poor maintenance in the energy sector are symptoms of a much deeper problem embedded in policy, institutional, and market failures for the development and management of energy resources in Pakistan and the region at large. The analytical framework is depicted in **Figure 2**. The most common divergence occurs due to policy failure often measured by the amount of direct or indirect subsidy provided. The marginal cost of production (MPC) representing the cost of producing the commodity, which is power in our case, shifts to the right due to the subsidy (reduction in cost) resulting in enhanced production up to Qo instead of Qi. The policymakers can internalize this cost (FI) by removing subsidies and other market distortions which have embedded themselves into the energy sector overtime.

The second reason for divergence between private and social costs is institutional failure. This is often reflected in poor governance and planning which is pervasive in across the energy sector in Pakistan. The cost of poor governance shifts the MPC to the left and is indicated by MPC plus marginal user cost (MUC), and the external cost is reflected by (EF). The economic instruments often used to internalize institutional failure include providing secure rights if there is no divergence between private and social discount rate. In case discount rate divergence exists, external costs can be corrected through output taxes and other economic instruments.

The last important component is environmental externality due to market failure (public good and natural monopoly) often referred to as marginal environmental or damage cost (MEC). It is argued that due to market failure, important social costs and benefits are left outside producers and consumers budgets or are one of those

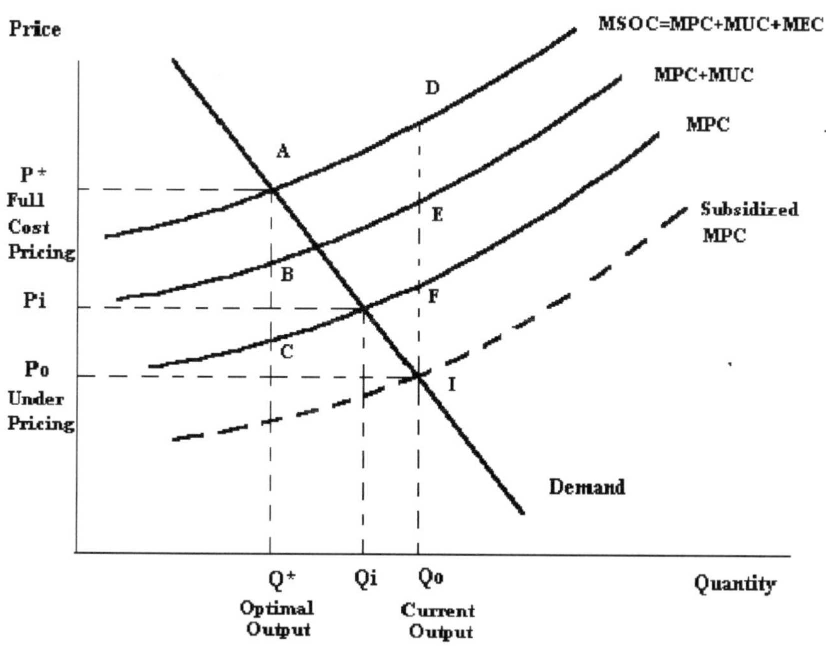

Figure 2.
Concept of policy, institutional, and market failure in energy sector.

unaccounted social costs (DE). The producer and consumer often perceive costs for environmental services as zero primarily due to lack of markets. The MEC is often internalized through taxes, charges, tradable permits, or other economic instruments.

4. Policy failure

Pakistan's energy sector is plagued by persistent policy failures, exacerbating the country's dependence on imported fossil fuels, despite abundant renewable energy potential. Misaligned demand forecasting, inefficient subsidies, and inadequate planning have resulted in a crippling circular debt of over PKR 2.6 trillion (approximately USD 9.3 billion) [6]. The sector's lack of cash flow, coupled with ineffective regulatory frameworks, hampers investment in domestic energy resources, particularly renewables. For instance, despite the presence of a dedicated "Alternate and Renewable Energy (ARE) policy" since 2019, which mandates the carrying out of competitive bidding for future renewable energy projects, the government has been unable to carry out auctions, leaving the future of 6–7 GW of wind and solar energy projects hanging in the balance [18].

Reliance on imported fuels drains Pakistan's foreign exchange reserves, while subsidies burden the economy, discouraging energy efficiency and private sector participation. The failure to transition toward a sustainable energy mix threatens energy security, hinders economic growth, and undermines efforts to mitigate climate change, emphasizing the urgent need for comprehensive policy reforms.

4.1 Misaligned demand

As highlighted above, Pakistan's energy sector has been a case of policy failures and misaligned demand forecasts. The country's installed capacity stands at 40,813 MW in 2024, significantly exceeding peak demand estimates of 28,252 MW (average 25,000 MW, with seasonal variations) [19]. However, transmission capacity constraints limit delivery to only 23,000 MW, creating a shortfall in meeting peak demand. Furthermore, distribution losses of 17% (13% as per NEPRA's determination) exacerbate supply-side inefficiencies, passing additional costs on to consumers [19].

These policy misalignments have severe fiscal implications, contributing to Pakistan's widening fiscal deficit and necessitating frequent financial bailouts.

4.2 Subsidy burden: A financial albatross

Pakistan's energy sector is plagued by unsustainable subsidies, distorting the market and burdening government finances. Energy tariffs in Pakistan are often not reflective of the actual cost of generation and distribution, particularly when oil and gas prices are high on the international market, in other words, tariffs are set below cost-recovery levels, resulting in financial losses and rising circular debt. Subsidized consumers often have little incentive to invest in energy efficiency and demand side management leading to continued distortion in demand.

In FY 2022, subsidies reached PKR 600 billion, with significant delays in reimbursement, exacerbating liquidity issues and circular debt. For FY 2023–2024, power sector subsidies are estimated at PKR 976 billion ($3.5 billion), including direct subsidies and circular debt financing [20].

4.2.1 Inefficient tariff structure and poorly targeted subsidies

The current tariff structure disproportionately benefits residential and agricultural consumers, with cross-subsidization from industrial and commercial users [21]:

- Residential consumers: PKR 24 to PKR 50 per unit (progressive slabs)
- Commercial consumers: PKR 39 to PKR 47 per unit
- Industrial consumers: PKR 32 to PKR 45 per unit
- Agricultural consumers: PKR 30 to PKR 41 per unit
- Fuel cost adjustments: Monthly adjustments based on fuel prices.

These subsidies often benefit a broader population instead of targeting vulnerable or low-income groups, perpetuating inefficiencies. The government often sets artificially low tariffs for protected residential and agricultural consumers, as a means of providing social welfare. While politically popular, this limits the ability of power companies to generate sufficient revenue to cover operating costs and reinvest in the sector.

Pakistan's electricity prices are set by the government regulator, NEPRA, rather than market forces, leading to inefficient pricing and several negative consequences. These include the under-recovery of costs, distorted incentives for producers, distributors, and consumers, and the government's inability to pass on high import-based energy costs to consumers. This has resulted in mounting circular debt, significant price distortions, and unaffordable energy for the public. Most critically, it has made Pakistan's industry uncompetitive. As highlighted in media reports and mentioned earlier, the cost per unit of electricity in Pakistan is Rs 60, compared to much lower rates in competitor countries: Rs 17 in India, Rs 18 in Nepal and Afghanistan, Rs 16 in Bangladesh, and Rs 10 in Iran, the lowest in the region [22].

These high electricity prices render Pakistani products uncompetitive against regional markets, cutting into industry profits and forcing industry closure in some cases.

4.3 Investment deterrent

The energy sector's financial instability, marked by circular debt, unreliable cash flows (discussed above), and delayed payments to IPPs, deters international investors, especially large innovative projects such wind and solar parks and floating energy that brings economies of scale and open possibilities for an integrated energy policy on one hand and huge investment gap in developing a sustainable grid and transmission lines to deliver energy where demanded [1, 2].

A case in point is the government's repeated attempts to float tenders for a 600 MW utility scale solar park in Muzaffargarh, which failed to attract even a single bidder. Investors cited concerns about the country's broader macroeconomic stability, the financial health of the off-taking utility, and the high-risk non-payment [15, 16].

More recently the government's forceful termination of contracts with 5 IPPs and on-going renegotiation with 18 others has also rattled investor confidence. Many foreign sponsors have taken the issue up with the government [23].

5. Institutional failure

Pakistan's persistent energy shortages can be attributed to profound institutional failures, manifesting in inefficient public sector operations, antiquated equipment, and outdated infrastructure. A good example is how state-owned enterprises like Pakistan State Oil (PSO), Pakistan Electric Power Company (PEPCO), and Pakistan Petroleum Limited (PPL) as entities are plagued by inefficiencies and corruption, bleeding the economy dry. The overall revenues of all state-owned enterprises in 2018–2019 were approximately Rs. 4 trillion, with assets valued at Rs. 19 trillion [24]. However, these enterprises are a huge burden on the public exchequer without delivering results, distorting competition and crowding out the private sector through inefficient operations and excessive commercial lending guaranteed by the government.

Despite attempts at reform, the energy generation landscape remains disproportionately controlled by the public sector, accounting for 52% of production. Decades of monopolistic dominance have stifled competition, perpetuating reliance on imported fossil fuels. Policy initiatives aimed at deregulating the sector and fostering competition have been hindered by entrenched interests, political resistance, and implementation delays, often orchestrated by vested groups within the public sector and private interests [24].

This section examines the entrenched institutional flaws undermining Pakistan's energy sector, highlighting the urgent need for structural overhaul and governance reforms to unlock a more efficient, diversified, and sustainable energy future.

5.1 Power theft and system losses

Power theft through illegal connections, known as "kunda" systems, exacerbates the problem, with estimated annual losses of $2.5 billion due to electricity theft and system losses [25]. This increases the financial burden on public sector companies, reducing their ability to generate revenue. Rent-seeking and political interference is marked by chronic corruption and undermine the sector's credibility, with contracts often awarded based on political patronage rather than merit. This alienates potential investors and hampers growth. Governments have failed to implement necessary structural reforms due to political expediency.

5.2 Absence of coordination among institutions

The energy regulatory agencies lack the capacity and autonomy to effectively regulate the sector. The failure of various public institutions to coordinate on energy production, transmission, and distribution further deepens the sector's inefficiency. The lack of an integrated energy planning approach, with conflicting interests and short-term goals, has led to institutional disarray. Reform is crucial to address these challenges to: enhance transparency and accountability in state-owned enterprises; implement cost-reflective tariffs to reduce subsidies and promote efficiency; promote private sector investment to increase competition and innovation and strengthen regulatory oversight to prevent corruption and ensure merit-based contracts.

5.3 Poor policy implementation

Governments have failed to follow through on necessary structural reforms due to political expediency. The commitment to privatization, restructuring, and tariff reforms is inconsistent, contributing to institutional inefficiency. Policies are often poorly implemented, and decisions are influenced by political considerations rather than technical expertise [26].

Pakistan's power transmission and distribution system is notorious for high technical and non-technical losses. These losses, which occur during the transportation of electricity from power plants to consumers, are among the highest in the region, ranging from 18% to 20% compared to the global average of about 8% [19]. The energy sector in Pakistan operates with a bizarre mix of public and private entities. While generation was privatized, transmission and distribution are largely under government control, creating an inefficient system with divided responsibility. This public-private mix leads to overlapping mandates and slow decision-making. While NEPRA is technically autonomous, it is still a state entity and is often subject to political interference. This has led to inconsistent regulatory decisions, delays in tariff adjustments, and a lack of transparent market mechanisms.

5.4 Lack of regulatory capacity

Lack of regulatory capacity: The energy regulatory agencies lack the capacity and autonomy to effectively regulate the sector. The failure of various public institutions to coordinate on energy production, transmission, and distribution further deepens the sector's inefficiency. The lack of an integrated energy planning approach, with conflicting interests and short-term goals, has led to institutional disarray.

5.5 Lack of innovation

As nations worldwide work to triple renewable energy capacity by 2030 and achieve net-zero emissions by mid-century, solar PV and wind power have emerged as cost-effective solutions. However, the variable nature of these renewables presents growing challenges for system management as their share in the energy mix increases. While chemical batteries address short-term storage needs, the critical gap lies in long-duration energy storage (LDES), with pumped hydro storage (PHS) standing out as the most established and scalable option. Recognizing the need to support LDES and PHS development, Pakistan's northern areas and the dessert terrain in the Thal region provide a good opportunity in adopting new technologies such as smart grids, renewable energy integration, and innovative energy storage as explained in **Box 1**.

> Farmers in Tehsil Mankera can significantly improve their living conditions by adopting solar tube well systems, which are more cost-effective compared to conventional tube well irrigation methods. To optimize this transition, efforts would be needed to modify sand dunes, store rainwater, and conserve both water and wildlife to adapt to changes in temperature and climate. Pumped Hydro Storage (PHS) presents a practical and cost-efficient solution to support modern agriculture in the region.
>
> The Thal Desert provides vast areas suitable for reservoir construction and offers natural elevation differences ideal for PHS. Mankera, located near the Jhelum River with an established canal network, is strategically positioned for such a system. Additionally, the area benefits from groundwater availability at economical depths, with good water quality. With an elevation ranging from 150 to 250 meters, Mankera provides sufficient head for PHS implementation. The scalability of PHS allows it to be adjusted based on energy storage requirements, making it an adaptable and sustainable approach for the region.

Potential benefits of PHS:

1. Renewable energy integration: can stabilize the grid and facilitate integration of solar and wind energy.
2. Peak demand management: can supply electricity during peak demand periods.
3. Energy security: provides energy storage for emergency situations.
4. Job creation: construction and operation can generate local employment.
5. Economic growth: can attract industries and investments to the region.

The proposal is not without challenges, first the high upfront cost which can be addressed by government incentives, private investment, or international funding can help. Secondly, environmental concerns, which can be addressed by conducting a thorough environmental impact assessments and implement mitigation measures and ensuring efficient water usage and management.

Source: Author field visit to Mankera region, 2024.

Box 1.
Pumped Hydro Storage (PHS) is a good option for the Mankera region of Thal Desert in Pakistan.

5.6 Aging infrastructure

Pakistan's energy sector is hindered by aging infrastructure, particularly in power plants. Most of the country's power generation capacity relies on outdated thermal power stations, many using technology that is no longer efficient or reliable. This results in low efficiency (below 30% compared to international standards of 40–45%), higher operational costs, frequent breakdowns, and difficulty maintaining a steady energy supply.

Financial constraints and ineffective oversight exacerbate the issue, leading to delayed or skipped maintenance cycles. Small technical issues escalate into major breakdowns, causing extended plant outages and energy shortages. For instance, the Guddu Thermal Power Station, built in the 1980s, has faced repeated breakdowns due to technical faults, resulting in substantial losses in electricity generation. In January 2021, an engineering fault at the Guddu power station tripped the entire grid plunging the whole country into darkness [27]. Just 2 years later in January 2023, the nation again suffered a blackout for over 12 hours due to a voltage surge [28].

To address these challenges, Pakistan needs to invest in modernizing the grid, implement regular maintenance schedules, enhance oversight and monitoring, and promote efficient and reliable energy generation. This requires a concerted effort to upgrade infrastructure, prioritize maintenance, and adopt new technologies to improve the sector's performance and meet growing energy demands.

5.7 Private-led solar energy boom

Pakistan's increasing adoption of solar PV under both grid-tied and off-grid mode is a testament to the economic competitiveness of renewable energy in the region. Pakistan's current net metered capacity stood at 3.2 GW as of September 2024 growing from just 1.1 GW in June 2023 [18]. As shown in **Figure 3**, most of this capacity (50%)

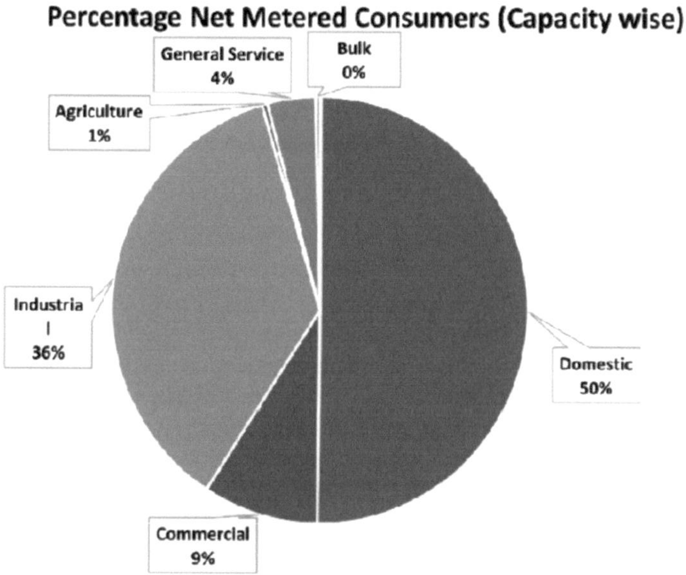

Figure 3.
Percent net metered (capacity wise) in Pakistan. Source: Faizan Ali [29].

has been put up by residential or domestic consumers while the rest can be attributed mainly attributed to the commercial and industrial (C&I) sectors.

This rapid increase in solar rooftop capacity has been facilitated by the drop in solar module prices and the extremely high cost of electricity in Pakistan. Average module selling prices dropped by 47% during fiscal year (FY) 2024, trending at USD 0.17 per watt by June 2024 (Muhammad [30]), while electricity prices rose by a 100% between FY 2022 and FY 2024. Combined with a favorable Net Metering Policy which allows consumers (up to 1 MW) to sell excess solar production to the grid at the prevailing National Average Power Purchase Price, net metered grid users can now break even between 2 and 4 years depending upon the system size and configuration [18].

Frequent grid outages and load shedding have also amplified consumer desire to move toward independent off-grid solutions. The same applies to the industrial sector as well, where critical manufacturing processes require uninterrupted power supply. In addition, the high cost of grid electricity for the industry at 17 US cents/KWh (without taxation) cuts into profit margins and reduces the regional competitiveness of Pakistani exports.

As a result, the industry has also been shifting away from the national grid toward decentralized solutions at an alarming pace.

Almost 12 GW of distributed solar PV is estimated to exist off-grid including single-phase non net metered residential consumers, Commercial and Industrial Installations, and agricultural tubewells. This rapid solarization of off-grid segments is projected to have direct implications for the future of the grid. Industrial grid demand dropped to 31 TWh in FY23, from 34 TWh in FY22 [30]. On a wider scale, power generation from the grid has been consistently dropping over the past 2 years, reducing by 10% in FY 23 [12] and 2% in FY 24 on a year-on-year basis [31]. As this trend continues into FY 25 [32], it may be rightly assumed that the grid demand has already peaked. A decrease in consumer demand from the grid paradoxically leads to a higher quantum of capacity payments, as the fixed nature of "take or pay" clauses force the government to pay for idle capacity.

5.8 Demand-side management needs to be prioritized

Pakistan's energy sector exhibits significant policy failures, particularly in demand management. Despite the potential for energy conservation and efficiency to alleviate supply pressures, demand management options are often overlooked in planning and policy formulation.

Effective demand management strategies can reduce overall energy demand, promoting a more efficient use of resources. Policies incentivizing energy-efficient technologies, such as: LED lighting, smart grid and meters, and energy audits can lead to significant savings and reduce the need for additional power generation capacity. This underscores the urgent need for policy reforms prioritizing demand management, energy efficiency, and market-driven pricing to ensure a sustainable and equitable energy sector that also look at credible storage options (see **Box 1**) [1, 2].

6. Market failure

Pakistan's energy market is plagued by entrenched market failures, distorting the efficient allocation of resources and hindering the transition to a sustainable energy mix. Information asymmetry pervades the sector, as consumers lack access to accurate data on energy pricing, quality, and supply, while investors face uncertainty regarding regulatory frameworks and project viability. Moreover, negative externalities, such as environmental degradation and health impacts from fossil fuel consumption, remain unpriced and unaccounted for, leading to overreliance on polluting energy sources.

Positive externalities, like the benefits of renewable energy adoption, similarly go uncompensated. This market failure section explores how inadequate pricing mechanisms, insufficient disclosure, and uninternalized externalities frustrate Pakistan's energy sector, necessitating targeted policy interventions to correct these market imperfections and promote a more efficient, equitable, and environmentally conscious energy market.

6.1 Information asymmetry

Consumers lack access to information about energy prices, quality, and availability. Consumers often face confusion regarding the breakdown of their energy bills, taxes, and subsidies. Many are unaware of the actual costs of electricity production, transmission, and distribution, which include capacity payments to power plants. The government frequently alters tariffs without clear communication, and consumers lack access to information on how rates are determined, leading to uncertainty.

6.2 Externalities ignored

The environmental and social impacts of energy production and consumption are neglected in policy decisions. Examples: Pakistan's energy sector subsidies amounted to PKR 1.4 trillion (approximately USD 9 billion) in 2020, straining the national budget [6]. The Pakistan State Oil (PSO) has been criticized for its inefficient operations, resulting in significant financial losses. This huge cost can be used to provide an upfront subsidy to renewable energy (solar and wind) to internalize policy failure but also mitigate environmental costs.

Market failures are seen in the monopoly control of transmission and distribution as well, distorted pricing mechanisms, and the absence of genuine competition, which together prevent efficient service delivery and adequate private sector investment.

Monopoly in transmission and distribution: Despite some privatization of power generation, transmission and distribution remain monopolized by state-run entities, which suffer from inefficiency, poor service, and outdated infrastructure. This creates bottlenecks in power delivery, frequent outages, and significant technical losses.

6.3 Lack of competition

The energy market lacks genuine competition, particularly in the distribution sector, where consumers are locked into inefficient, government-run distribution companies. Consumers are left with little choice, and this limits the incentives for these companies to improve service and reduce costs.

6.4 Negative environmental externalities

The failure to diversify the energy mix has led to Pakistan's over-reliance on fossil fuels, particularly imported oil and coal, contributing to environmental degradation. Market mechanisms have failed to incorporate these external costs, such as pollution and greenhouse gas emissions, into energy prices.

There is little incentive for private investment in the energy sector due to the high regulatory and political risk. The market is not functioning effectively because investors cannot trust the stability of policy, and the government's frequent intervention in pricing and contracts erodes confidence in market dynamics.

7. Renewable energy: Policy path by India in comparison with Pakistan

A few decades ago, India's energy sector, like Pakistan, could be characterized as a policy and institutional failure, but thanks to central government policies and initiatives technological advancements and significant foreign investments India has witnessed remarkable growth in the renewable energy capacity. The scale of transformation in India is stunning; India's economic growth has been among the highest in the world lifting millions of people out of poverty every year. This energy transformation is evident in the growth achieved in renewable energy with commitment to achieve 500 GW of by 2030. India is on track to set a record in renewable energy capacity—to connect 35 GW of solar and wind power to its grid by March 2025, the world's third largest solar power producing nation is ramping up its efforts to meet the ambitious renewable energy [33]. India also aims to produce 5 million tons of green hydrogen by 2030. It further aims to reach net zero emissions by 2070, and to meet fifty percent of its electricity requirements from renewable energy sources by 2030 is a hugely significant moment for the global fight against climate change [34].

Pakistan and India, two neighboring South Asian nations, have both recognized the importance of renewable energy to address their energy shortages, enhance sustainability, and mitigate climate change. However, the approaches they have adopted in their renewable energy policies differ in several ways. Pakistan's renewable energy policies have evolved slowly compared to India's. Pakistan introduced its Alternative and Renewable Energy (ARE) Policy in 2019 [35], focusing on increasing the share of renewables like wind, solar, and hydro. The policy sets a target of achieving 30% of energy from renewable sources by 2030, complementing hydropower which is already a major contributor to the country's energy mix. In contrast, India began its renewable energy push much earlier, with aggressive policy frameworks that target 50% of electricity generation capacity from non-fossil sources by 2030, driven by solar and wind [36]. These targets have been progressively revised and stood at an impressive 500 GW of non-fossil fuel based capacity by 2030 [37].

A key difference between Pakistan and India in exploring renewable energy potential lies in institutional frameworks and financing mechanisms. India benefits from a robust institutional infrastructure, led by the Ministry of New and Renewable Energy (MNRE) and supported by state-level agencies that actively promote decentralized renewable energy initiatives. Mechanisms such as competitive auctions for large-scale solar projects and incentives for private sector participation have propelled India's progress. For instance, Rajasthan's abundant sunlight and vast desert tracts have facilitated the development of iconic projects like the Bhadla Solar Park, one of the world's largest solar parks. In contrast, similar tracts on Pakistan's side of the border remain untapped, where a coal-based power plant has been developed instead (see **Box 2** for details).

Rajasthan, India, has emerged as a leader in renewable energy development, with its success largely attributed to its strategic use of solar and wind resources. The state's abundant sunlight and vast tracts of desert land have made it ideal for solar power projects, including large-scale solar parks like the Bhadla Solar Park, one of the world's largest. Through policy support, investment incentives, and public-private partnerships, Rajasthan has attracted significant domestic and foreign investments. The government also simplified regulatory frameworks and improved grid infrastructure to ensure efficient energy transmission, helping scale up renewable energy installations. Pakistan can draw valuable lessons from Rajasthan's approach to renewable energy. First, aligning national policies to encourage investments in renewables and simplifying the regulatory process can significantly reduce the cost and time needed to deploy renewable projects. Pakistan's potential in solar and wind energy, particularly in regions like Tharparker, Sindh and vast tracts in Baluchistan, could be harnessed similarly by identifying optimal locations and facilitating infrastructure development. Furthermore, fostering partnerships between the public and private sectors can attract the necessary financial and technical expertise. Moreover, Rajasthan's efforts in community engagement and employment creation offer insights into the social benefits of renewable energy projects. By involving local communities in project planning and providing jobs in both construction and operations, Rajasthan mitigated social resistance and enhanced the local economy. Pakistan could replicate this model by ensuring that renewable energy projects involve and benefit local populations, especially in rural areas, helping to achieve both energy independence and socioeconomic development.

Opportunities for Pakistan:

1. Learning from India's success stories

2. Improving infrastructure and transmission lines

3. Enhancing policy frameworks and incentives

4. Encouraging private sector investment.

Source: Rajasthan Renewable Energy Corporation Limited (RRECL).

Box 2.
Rajasthan India and Thar Pakistan—Renewable energy paths.

Pakistan faces significant institutional and policy challenges, including limited capacity for project implementation and minimal private sector involvement. Its heavy reliance on donor-funded projects has created bottlenecks, exacerbated by underdeveloped domestic financing structures. By 2024, India had achieved over 205 GW of renewable energy capacity [38], led by initiatives like the National Solar Mission, which aggressively expanded solar and wind energy through falling costs, international investments, and large-scale renewable energy parks. In contrast, Pakistan's renewable energy capacity remains below 10 GW, hindered by political instability, subsidies for fossil fuels, and bureaucratic delays. Despite its vast solar potential, Pakistan has yet to leverage these resources effectively, trailing significantly behind India's renewable energy progress [39].

Key lessons from the comparison include the importance of a robust policy framework, strong institutional backing, and conducive financing mechanisms for fostering renewable energy growth. Pakistan could learn from India's decentralized approach to renewable energy, enabling state governments and local bodies to implement region-specific solutions. India's Green Energy Corridor, launched in 2012 by the Power Grid Corporation of India Limited (PGCIL), is a landmark transmission infrastructure project designed to seamlessly integrate solar and wind energy into the national grid [40], addressing the challenge of connecting renewable energy generation sites with demand centers (explained further in **Box 3**). Further the success of India's auction mechanisms and incentives to lower costs is another key takeaway. Additionally, Pakistan could benefit from reducing its reliance on external funding by developing domestic financing mechanisms and partnerships to accelerate renewable energy projects.

India's Green Energy Corridor is a transmission infrastructure project that aims to integrate renewable energy sources, particularly solar and wind power, into the national grid. The project was initiated in 2012 by the Power Grid Corporation of India Limited (PGCIL) to address the challenge of transmitting renewable energy from generation sources to demand centers.

Key objectives:

- Evacuation of renewable energy: Create a dedicated transmission infrastructure to facilitate the smooth transfer of renewable energy from generation sources to demand centers.
- Grid stability: Enhance the stability and reliability of the electricity grid to accommodate intermittent renewable energy sources.
- Reduced carbon emissions: Contribute to a reduction in carbon emissions by promoting clean energy transmission.

Phases of development:

The Green Energy Corridor project is being implemented in phases. Phase 1, which began in 2015, focuses on eight renewable-rich states: Gujarat, Andhra Pradesh, Karnataka, Himachal Pradesh, Maharashtra, Madhya Pradesh, Tamil Nadu, and Rajasthan. Phase 2, approved in January 2022, will facilitate the grid integration and power evacuation of approximately 20 GW of renewable energy in seven states.

https://powermin.gov.in/en/content/green-energy-corridor.

Box 3.
Green Energy Corridor in India.

8. Conclusion: Favorable pricing and economic necessity win over system inefficiencies and market forces

Pakistan's energy sector exemplifies the intersection of policy, institutional, and market failures:

Policy failures stem from inconsistent reforms, short-term political objectives, and a lack of strategic foresight, especially in pricing, privatization, long-run exchange rate misalignment, and long-term investment planning.

Institutional failures are evident in weak governance, fragmented responsibilities, and an inability to coordinate among various public and private players, compounded by corruption and political interference.

Market failures are seen in the monopoly control of transmission and distribution, distorted pricing mechanisms, and the absence of genuine competition, which together prevent efficient service delivery and adequate private sector investment.

The cumulative impact of these failures is an energy sector that is expensive, inefficient, and unable to meet the country's growing energy needs. The sector's dysfunction is reflected in rising circular debt, unaffordable tariffs, frequent power outages, and a growing reliance on imported energy—leaving the country vulnerable to external shocks and threatening economic stability.

Although positive progress has been observed through a rapid increase in rooftop solar capacity, this comes with its own set of grid-related challenges if left unchecked. An increase in off-grid renewable energy capacity may be desirable from an energy independence point of view, empowering consumers to choose their own electricity supply; however, this could increase the economic burden of maintaining the national grid and existing power capacity in the country on grid-connected consumers.

To rectify the situation, the government needs to urgently invest in grid modernization and advanced monitoring of power flows to and from the grid. More importantly, demand creation is of utmost importance to stem the exit of large industrial consumers from the grid.

Pakistan could also benefit by adopting India's decentralized approach, empowering local bodies to tailor solutions according to regional needs. Learning from India's successful use of auction mechanisms and incentives to reduce costs, Pakistan could also advance by establishing domestic financing options and partnerships to lessen reliance on external funding for renewable energy projects.

Author details

Mahmood Ahmad[1]*, Haneea Isaad[2] and Furqan Ali Akhtar[3]

1 YUNUS-RIZQ Knowledge Hub (Pakistan, Thailand) Yunus Centre, Asian Institute of Technology (AIT), Bangkok, Thailand

2 Institute for Energy Economics and Financial Analysis (IEEFA), USA

3 Pantera Energy, Lahore, Pakistan

*Address all correspondence to: mahmoodahmad@ait.asia

References

[1] IEA. Energy Efficiency Report. 2022

[2] IEA. World Energy Outlook 2022—Analysis—IEA. France: International Energy Agency; 2022. Available from: www.iea.org/reports/world-energy-outlook-2022

[3] Hasan M. Energy economic expansion with production and consumption in BRICS countries. Energy Strategy Reviews. 2022;**44**:101005. DOI: 10.1016/j.esr.2022.101005 [Accessed: November 22, 2022]

[4] China Briefing. China's new renewable energy plan: Key insights for businesses. China Briefing News. 2024. Available from: www.china-briefing.com/news/chinas-new-renewable-energy-plan-key-insights-for-businesses/

[5] Archana et al. Development of smart grid for the power sector in India. Cleaner Energy Systems. 2022;**2**:100011. Available from: www.sciencedirect.com/science/article/pii/S2772783122000103. DOI: 10.1016/j.cles.2022.100011

[6] World Bank Group. Pakistan Development Update; The Dynamics of Power Sector Distribution Reforms. Pakistan: World Bank Group; 2024

[7] Pakistan Bureau of Statistics. Government of Pakistan Pakistan Bureau of Statistics. Pakistan: Pakistan Bureau of Statistics; 2023

[8] Government of Pakistan. Updated Nationally Determined Contributions 2021. Pakistan: Ministry of Climate Change, Government of Pakistan; 2021

[9] World Wildlife Fund. Nature-Based Solutions Termed Critical to Address Climate Change Impacts: Experts. 2024. Available from: www.wwfpak.org/?387681/Nature-based-solutions-termed-critical-to-address-climate-change-impacts-Experts [Accessed: December 26, 2024]

[10] Deutsche Gesellschaft für Internationale Zusammenarbeit (GIZ) GmbH. Digitisation and Decarbonisation of Power Distribution Networks in Pakistan. 2024. Available from: www.giz.de/en/downloads/giz2024-en-factsheet-decarbonisation.pdf

[11] Ghumman M. Nepra Says "No" to Economic Load Management. Pakistan: Brecorder; 2024. Available from: www.brecorder.com/news/40329499 [Accessed: December 27, 2024]

[12] Riaz N. The Double Whammy of Declining Electricity Demand and Increased Consumer Tariffs. Pakistan: Profit by Pakistan Today; 2024. Available from: https://profit.pakistantoday.com.pk/2024/07/29/the-double-whammy-of-declining-electricity-demand-and-increased-consumer-tariffs/ [Accessed: December 27, 2024]

[13] Ul-Haq A et al. Implementation of smart grid technologies in Pakistan under CPEC Project: Technical and policy implications. IEEE Access. 2021;**9**:61594-61610. DOI: 10.1109/access.2021.3074338 [Accessed: January 6, 2023]

[14] Shahid A. Pakistan Ends Power Deals to Save $1.48 Billion, Cut Tariffs. Pakistan: Reuters; 2024. Available from: www.reuters.com/business/energy/pakistans-biggest-private-utility-says-govt-power-deal-ends-prematurely-2024-10-10/

[15] Kiani K. Energy Sector Circular Debt Surges to Rs5.5tr. 2024. Available from:

www.dawn.com/news/1825571 [Accessed: December 27, 2024]

[16] Kiani K. Govt Fails to Woo a Single Investor for Solar Project. 2024. Available from: www.dawn.com/news/1805138 [Accessed: December 29, 2024]

[17] State Bank of Pakistan. Annual Report 2022-2023, the State of Pakistan's Economy. 2023. Available from: www.sbp.org.pk/reports/annual/aarFY23/Title.pdf

[18] Isaad H, Ali SF. The Future of Net-Metered Solar Power in Pakistan. Pakistan: Institute for Energy Economics and Financial Analysis; 2024

[19] National Electricity and Power Regulatory Authority (NEPRA). NEPRA | Home. 2024. Available from: https://nepra.org.pk/publications/State%20of%20Industry%20Reports/State%20of%20Industry%20Report%202023.pdf

[20] Malik A, Mustafa G. PIDE Working Papers Power Sector Debt and Pakistan's Economy. Pakistan: Pakistan Institute of Development Economics; 2024

[21] Islamabad Electric Supply Company. Tariff Guide. 2024. Available from: https://iesco.com.pk/index.php/customer-services/tariff-guide

[22] Dunya News. Good News for Electricity Consumers-CPEC-Pak China Friendship-IPPs | Gohar Ejaz Interview-Dunya News. Pakistan: YouTube; 2024. Available from: www.youtube.com/watch?v=k1-tDRpst00 [Accessed: December 29, 2024]

[23] Profit. Germany Objects to Terms of Pakistan's IPP Agreements, Cites Concerns over Deal with Rousch Power. Pakistan: Profit by Pakistan Today; 2024. Available from: https://profit.pakistantoday.com.pk/2024/10/30/germany-objects-to-terms-of-pakistans-ipp-agreements-cites-concerns-over-deal-with-rousch-power/ [Accessed: December 29, 2024]

[24] Iftikhar MN. State Owned Enterprises in Pakistan Prime Policy Report. Pakistan: Prime Institute; 2017

[25] Ministry of Energy, Power Division. Power Sector Theft. Pakistan; 2024

[26] Malik A. Privatisation of Electricity Distribution Companies—A Way Forward? Pakistan: Pakistan Institute of Development Economics; 2022

[27] Ali SA. Power Blackout 2021: Inquiry Report. Pakistan: Brecorder; 2022. Available from: www.brecorder.com/news/40172455 [Accessed: December 30, 2024]

[28] Hussain A. Pakistan Hit by Nationwide Power Outage after Grid Failure. 2023. Available from: www.aljazeera.com/news/2023/1/23/pakistan-hit-by-nationwide-power-outage-after-grid

[29] Shah FA. Solar Showdown: The Untold Power Play behind Pakistan's Energy Revolution—I. Pakistan: Brecorder; 2024. Available from: www.brecorder.com/news/40329650/solar-showdown-the-untold-power-play-behind-pakistans-energy-revolutioni [Accessed: December 30, 2024]

[30] Mustafa M et al. The Great Solar Rush in Pakistan the Great Solar Rush in Pakistan. Pakistan: Renewables First; 2024

[31] Business Recorder. FY24 Power Generation Drops 2pc YoY. Pakistan: Brecorder; 2024. Available from: www.brecorder.com/news/40313921/fy24-power-generation-drops-2pc-yoy [Accessed: December 30, 2024]

[32] Siddiqui S. Power demand slumps 17 in Aug. The Express Tribune. 2024. Available from: https://tribune.com.pk/story/2497324/power-demand-slumps-17-in-aug [Accessed: December 30, 2024]

[33] ANI. India's Renewable Energy Capacity Grew by 165% in 10 Years: Pralhad Joshi. India: @Bsindia, Business Standard; 2024. Available from: https://www.business-standard.com/india-news/india-s-renewable-energy-capacity-grew-by-165-in-10-years-pralhad-joshi-124080700090_1.html [Accessed: December 30, 2024]

[34] Bradstock F. India's Green Transition Gains Momentum with $386 Billion in Funding Pledges. 2024. Available from: https://oilprice.com/Energy/Energy-General/Indias-Green-Transition-Gains-Momentum-with-386-Billion-in-Funding-Pledges.html [Accessed: December 30, 2024]

[35] Government of Pakistan, Ministry of Energy. Pakistan introduced its Alternative and Renewable Energy (ARE) Policy in 2019

[36] Government of India, Press Information Bureau. Year End Review—Solar Power Target Reset to One Lakh MW. 2015. Available from: https://pib.gov.in/newsite/printrelease.aspx?relid=133220

[37] Government of India, Ministry of Power. 500GW Nonfossil Fuel Target | Government of India | Ministry of Power. 2024. Available from: https://powermin.gov.in/en/content/500gw-nonfossil-fuel-target

[38] Mittal V. India's Renewable Energy Growth in 2024—Et Edge Insights. India: ET Edge Insights; 2024. Available from: https://etedge-insights.com/industry/energy/indias-renewable-energy-growth-in-2024/ [Accessed: December 30, 2024]

[39] Global Petrol Prices. India Electricity Prices, 2021. 2021. Available from: www.globalpetrolprices.com/India/electricity_prices/

[40] KfW Development Bank. Green Energy Corridors Deliver Electricity. Germany: KfW Development Bank; 2022. Available from: https://www.kfw-entwicklungsbank.de/SDG-portal/SDG-13/Green-energy-corridors-India/

Chapter 7

Can Eco-Innovation Ensure Sustainable Development? Unlocking Renewable Energy in Turkey

Ziyafat Guluzade, Anna Montini and Rui Alexandre Castanho

Abstract

Energy is a significant factor in nature and social life. Renewable energy sources have come to the fore in recent years due to their local quality, low fuel cost, and clean and environmentally friendly properties. In Turkey, where a significant increase in energy consumption has been observed in recent years, the increasing energy demand has been met, and the process of meeting this demand is being managed with sustainable energy policies. Turkey's renewable energy policies and the reasons for the rapid increase in the trend toward renewable energy sources are noted and evaluated in terms of their social, economic, and environmental impacts. The main research questions addressed how policies promoting renewable energy align with Turkey's commitment to environmental sustainability and contribute to the country's overall goals for sustainable development. What are effective communication strategies for socially accepting environmentally friendly renewable energy policies in Turkey and their dissemination throughout the country? This research will use a survey methodology to investigate the complex interplay between renewable energy adoption and environmentally friendly policies and their impact on sustainable development in Turkey. This survey methodology aims to capture a nuanced understanding of public perspectives that can contribute to Turkey's ongoing discourse on sustainable development.

Keywords: renewable energy, sustainability, environment, energy policies, environmentally friendly policies

1. Introduction

Today, there are many sources of energy. The primary source of energy in nature is the sun. Plants receive direct solar energy and convert it into chemical energy through photosynthesis. Energy in nature can be found in many different forms, such as mechanical, chemical, nuclear, electrical, thermal, and light energies, and each type of energy can be converted into another energy.

The Republic of Turkey has made great strides in recent years. However, the commercialization of low-calorie coal in a clean way, the negative impact of expensive energy prices on the current account deficit and, therefore, the exchange rate, a long-term hydrogen road map, and a clear sustainable energy strategy have been identified as the most critical issues to be overcome quickly on the way to 2030, for the benefit of decision-makers and other researchers.

In this research, the sustainable energy policies of the Republic of Turkey will be examined in light of information obtained from official and international sources, and the sustainability of these policies will be discussed in terms of environment and energy efficiency. In the conclusion part, determinations will be made through the evaluation of the data. With these determinations, it will try to shed light on what policies the Republic of Turkey has achieved in the past 10 years and what position it will be in by 2030.

The methodology of this research consists of literature analysis and questionnaire research. The literature review covers the impact of renewable energy on sustainable development and has been analyzed in detail. In addition, various policy documents and strategic plans were examined, and Turkey's energy policies' current state and future goals were evaluated.

Questionnaire research was conducted to popularize renewable energy use and assess the public's general opinion and awareness of renewable energy sources and environmental sustainability.

In fact, this methodological approach offers a comprehensive and integrated analysis to assess the impacts of renewable energy integration in Turkey on sustainable development and environmental sustainability and to develop specific recommendations in this area.

2. Conceptual framework

2.1 Main goals and objectives of sustainable development

The concept of sustainable development, which came up frequently in the late 1970s, envisages the use of resources in any field in a way that does not prevent meeting the needs of the communities that will exist in the world. Contextually, the future and the ability of these resources to renew themselves while maintaining their existence in the long term. According to this definition, the emerging sustainable development approach is expressed as preventing the overuse of natural resources and the environment, considering future generations, and allowing existing resources to renew themselves [1].

In September 2015, the United Nations organized the Sustainable Development Summit to assess 15 years of advancement toward the Millennium Development Goals set in 2000. The review concludes that attempts to attain this aim have yet to be well short of projected outcomes in resolving the world's economic, environmental, and social concerns. Recognizing the lessons learned from the Millennium Development Goals and the need to address existing challenges, the Summit emphasized the need to formulate new and more comprehensive sustainable development goals. In response to this call, between April 25 and 27, 2015, at the United Nations Headquarters in New York, 193 countries, including Turkey, adopted the Sustainable Development Goals until 2030. In this regard, it consists of 17 main goals and 169 targets; these goals aim to eliminate poverty in all dimensions and improve human well-being by 2030 [2].

The 17 main goals of the 2030 SDGs are called the new goals of the global sustainable development agenda [3]. Among these goals, SDG 7 (Affordable and Clean Energy) aims to provide access to modern and sustainable energy for all. Turkey has committed to supporting SDG 7 by transforming its energy systems and increasing the use of renewable energy sources. By 2030, it aims to increase the use of renewable energy by reducing the consumption of fossil fuels and doubling the global improvement in energy efficiency [4].

2.2 Sustainable development and energy relationship

Economic and social development is directly related to the country's development level. Energy represents one of the most significant components in achieving this level of progress. The Sustainable Development Goals promote a more sustainable and promising future for all people worldwide. Consequently, sustainability must be used to determine the ideal balance between energy production, environmental protection, and growth. One of the critical indicators to consider in terms of sustainable development is energy production methods [5]. Later, concerns about energy's economic, environmental, technical, and social benefits grew, and the issue was given great importance both locally and globally. Because all energy production systems, especially those that use fossil resources, harm the environment and ecosystem [6].

The focus areas of sustainable energy are energy security, resource management, and environmental protection. Five essential goals for a sustainable energy perspective are as follows: net zero carbon dioxide emissions, the absence of significant environmental harm, increasing the security of the energy transition, reducing the cost of energy, improving energy production, and using green technologies.

Although nature can tolerate small increases in carbon dioxide emissions, low-carbon technologies are critical to sustainable energy roadmaps, as high carbon dioxide emissions cause global warming effects. In addition, the sustainable energy perspective aims to manage energy flows and remove waste without exceeding nature's carrying capacity [7].

The seventh Sustainable Development Goal integrates affordable access to reliable, sustainable, and modern energy services for all. At a time when more than 80% of the world's consumption is based on fossil fuels, by 2030, it is aimed to significantly increase the share of renewable energy in the global energy mix and to double the rate of global improvement in energy efficiency [8]. Global climate change is a pressing issue in the twenty-first century due to high levels of greenhouse gases in the atmosphere, mainly CO_2 emitted from anthropogenic activities such as burning fossil fuels and deforestation. Continued increases in CO_2 emissions are expected to have significant consequences for the global climate system, with devastating consequences affecting all segments of society. As a result, elements such as reducing CO_2 emissions, renewable energy, and technical innovation have become important targets for attention to building a greener and more sustainable world. The benefits of renewable energy are that it reduces demand for conventional energy while maintaining long-term global economic output. Renewable energy is critical to achieving the global emissions reduction target of 50% by 2050 [9]. Renewable energy sources are recognized as among the most effective and cost-effective alternatives. Therefore, there is a strong relationship between renewable energy and sustainable development [10].

3. Turkey energy atlas: Overview, status, and potential of the renewable energy sector in Turkey

3.1 General evaluation of the energy sector in Turkey

The electricity sector is developing rapidly in Turkey and other developing countries. Population growth and growth rate are increasing in Turkey. The demand for power increases as industry and urbanization expand. The increase in living standards and economic growth in Turkey increased the need for electricity [10]. In 2001, with the transition of the electricity market to a free competition market, it was divided into three separate economic state enterprises: TEAS, TEIAS, and TETAS [11].

About half of the electricity produced in 2019 was produced in thermal power plants from fossil fuels such as natural gas and coal. Although renewable sources such as wind, solar, and geothermal have come to the fore in recent years, most production continues with fossil fuels. At this point, foreign dependence on fossil fuels and energy comes to the fore. Turkey's leading fossil energy source is coal, which has been produced locally for years and accounts for about 24% of the country's total energy consumption. It is mainly used for power generation, cement, and steel production. Turkey is one of the world's largest lignite producers. Therefore, this mainly comes from the South-West and South-East Afshin-Elbistan basin deposits, where 7339 million tons of lignite are economically available (**Figure 1**) [12].

According to **Figure 2**, Turkey's total primary energy supply in 2022 decreased by 1% compared to the previous year to 157.8 mtoe. When the total energy supply is examined based on resources, the share of fossil fuels in 2022 is 82.8%, and the share of renewable energy sources is 17.2%. Renewable energy sources will provide 27.1.

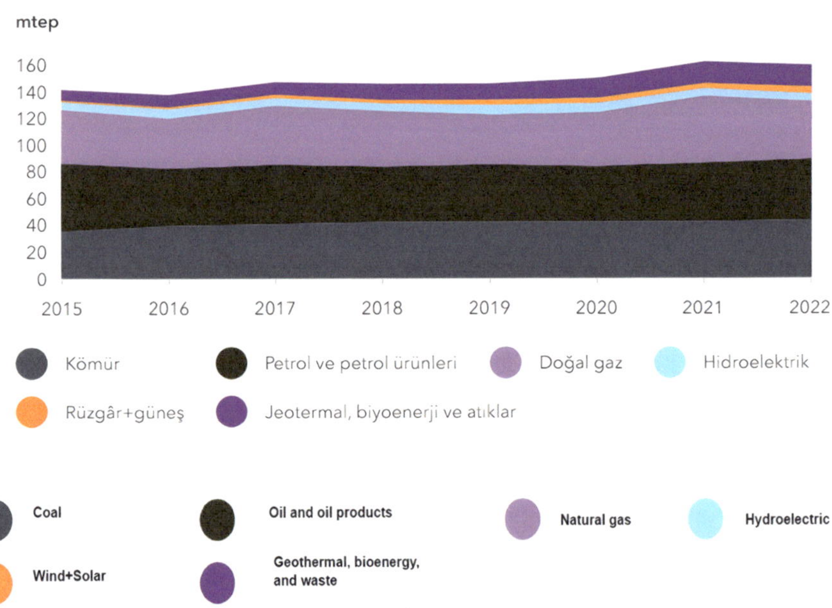

Figure 1.
Primary energy supply by source between 2015 and 2022 development [14].

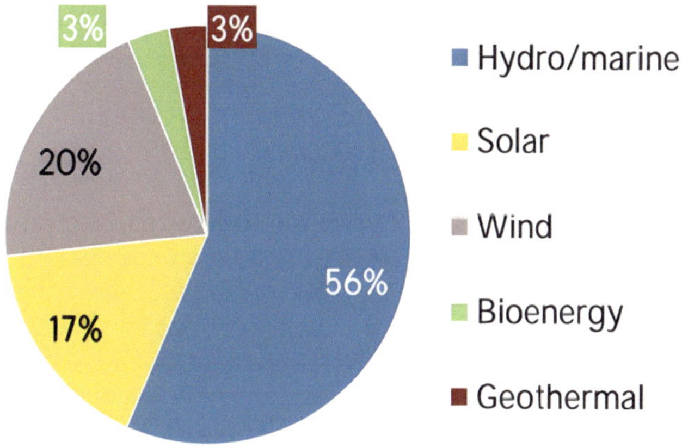

Figure 2.
Renewable energy capacity 2022 [14].

Moreover, several other sources are recommended to obtain a deeper knowledge on this issue – see [13–17]. Million tons to the primary energy supply by 2022, reaching the highest level since 2000. Despite this growth, the share of renewable energy in the total energy supply is growing relatively slowly (see: https://enerji.gov.tr/eigm-raporlari).

3.2 General overview and potential of renewable energy sources in Turkey

Turkey's geographical position provides a significant opportunity to develop renewable energy resources. Although it has high potential, energy production is low. This is because there are many factors, such as high installation costs of power plants and lack of legal regulations. Considering Turkey's foreign dependence on energy, producing renewable energy potential is essential for Turkey in the long term (**Figures 2** and **3**) [16].

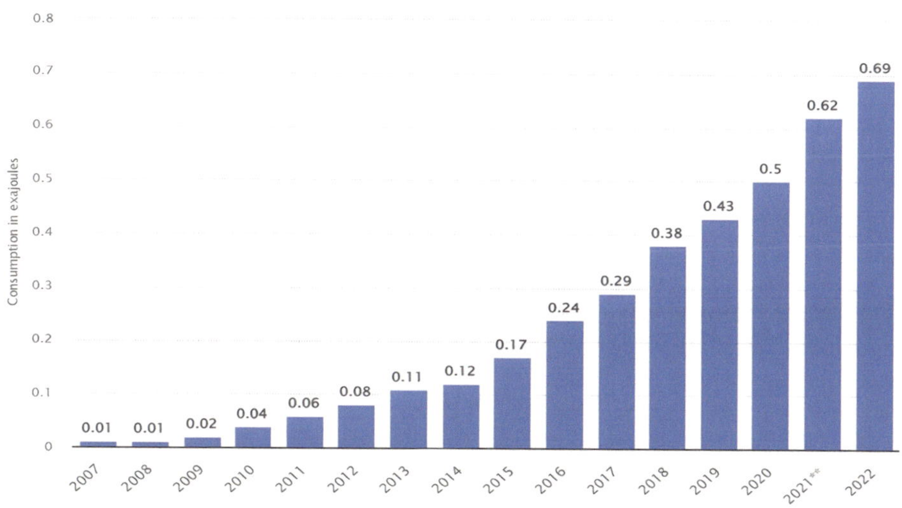

Figure 3.
Renewable energy consumption in Turkey from 2007 to 2022 [14].

Renewable energy consumption in Turkey has been increasing every year since 2008. Total consumption peaked in the observed period in 2022 and amounted to 690 petajoules.

3.2.1 Solar energy potential and utilization in Turkey

Due to its geographical location, Turkey is a country with four seasons and high solar potential. Turkey's first solar power plant was built in Konya and had a capacity of 200 KW. The number of installed solar power plants in Turkey is 556. The installed solar energy capacity of Turkey in 2018 was 5095 MW. Turkey is ranked 12th in the world for installed electricity. China ranks first with 131,000 MW of installed capacity, and Latvia ranks last with 2 MW. Turkey ranks 28th with 62 W of installed solar power per capita. According to the number of solar power plants per capita, Germany ranks first, with a capacity of 550 W [18].

3.2.1.1 Solar energy application examples in Turkey

The first application in our country for "Building Integrated Photovoltaic Systems" (BIPV), which is becoming more and more critical in PVPS applications, was launched in 2003 with the DPT project on the 215 m^2 south facade of the roof of the student cafeteria at Mugla University Campus. A grid-connected BIPV system has a total capacity of 25.6 kW, an average of 35,000 kWh of electricity, installed annually.

3.2.1.2 Pamukkale University "Clean Energy House Project"

The heating needs of "Pamukkale University Clean Energy House" come from solar energy; a solar-hydrogen hybrid system also provides electricity. Active and passive applications of solar energy are used for heating needs. The panels on which these batteries are installed follow the sun in the east-west and north-south directions throughout the year. Volther, produced by Solimpeks I.C., operating in Konya with 100% Turkish investment, was realized in a villa in Antalya/Kumluca. The application addresses all of the villa's power and hot water demands. In a PV-T hybrid system, cells are cooled by transferring waste heat to water, producing more efficient electricity and hot water for pool or house heating.

3.2.2 Wind energy

In this regard, Turkey's wind energy potential is defined as 48,000 MW. It corresponds to 1.30% of the territory of Turkey [15]. In 1998, Turkey developed its first wind power facility. This power station is the Çeşme-Germiyan Wind Power Station. The wind farm has only three turbines, and the power it produces is 580 kW. Since that date, the installation of wind power plants has increased; today, the total number of wind power plants is 180. While 172 wind power facilities provide energy, eight are under development (**Figures 4** and **5**). There are certain conditions for establishing a wind power plant in Turkey. These are: It is 50 m above ground level, and the wind speed is 7.5 m/s. Wind power plants with a capacity of 5 MW per square meter can be created in areas that meet these two conditions. For these power plants to be licensed, the wind flow must be measured for at least 1 year and must continuously be 7 m/s or above [15].

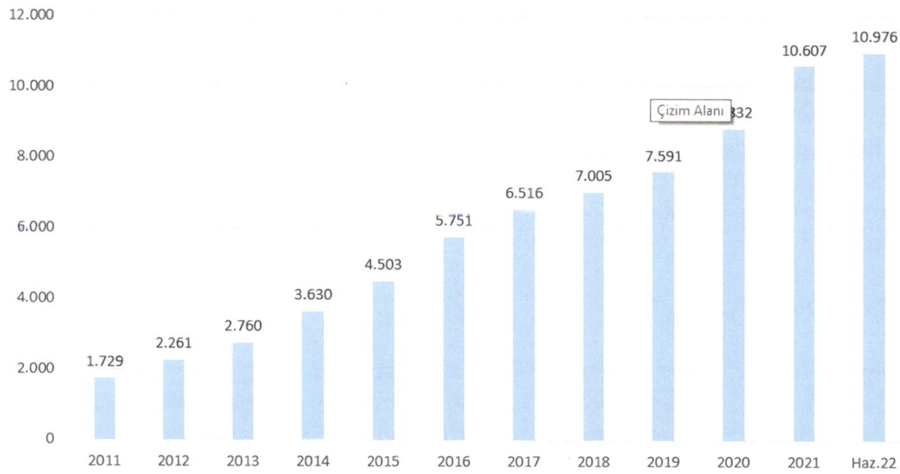

Figure 4.
Installed power based on wind energy (MW) [15].

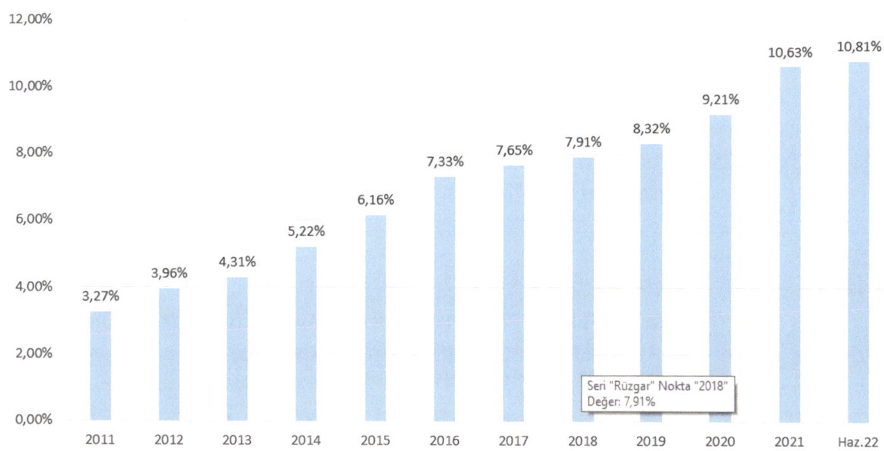

Figure 5.
Ratio in total installed power [15].

As of the end of June 2022, Turkey's electrical installed power based on wind energy is 10,976 MW, its ratio in total installed power is 10.81%, and the change in installed power by years and its ratio in total installed power are shown in the graphs above.

3.2.3 Geothermal energy

Turkey is one of the world's wealthiest countries in terms of geothermal resources because of its geologic and geographical position. There are 1000 geothermal resources in Turkey at different temperatures. Nine hundred of these geothermal resources have low to medium temperatures—heat, thermal tourism, mining, and fish farms are just a few examples. When we evaluate the geographical distribution of these geothermal resources, 78% are in Western Anatolia, 9% in Central Anatolia, 7% in the Marmara Region, 5% in Eastern Anatolia, and the remaining 1% in other regions (www.enerji.gov.tr/tr). The first electricity production from geothermal energy in our country was the Kızıldere Power Station, established by Mineral Research and Exploration (MTA) in 1975. About 90% of the geothermal energy available has low to medium temperatures and is appropriate for direct usage (warming, thermal tourism, and many other industrial applications), and 10% for indirect applications. Geothermal energy investments have increased over the years. There are 39 geothermal power plants in Turkey. Their total installed capacity is 1028.00 MW (**Figure 6**).

3.2.4 Hydraulic energy

Hydroelectric power plants (HEP) come first when meeting renewable energy needs in Turkey. When we look at the development of hydroelectric power in Turkey, the first hydroelectric power station built in 1929 consisted of two turbines. It produces 0.52 kW of power per turbine, and the company's name is Işıklar Water Power Station. In 2018, the number of hydropower plants was 624. The average annual electricity production by 624 companies is 72,198 GW/h. When comparing the production of hydropower plants and the total consumption of Turkey, the average consumption coverage value between 1970 and 2015 is 33.12%. Between 1970 and 2015, the highest indicator was 60% in 1988, and the lowest was 15.8% in 2014 [18].

In recent years, the demand for hydroelectric power has changed. Despite short-term volatility, hydropower production is increasing in the long term. The government's total installed capacity in 2023 is 32,000 MW. DSI aims to reach 127.8 TWh by 2030 [19, 20] and estimates that hydropower generation will increase by 30% by 2050.

3.2.5 Biomass energy

Turkey is an agricultural country, so there is a lot of agricultural and crop waste. Considering Turkey's biomass energy potential, product waste has a high potential with 9.5 million tons of oil equivalent (Mtoe).

Bioenergy Turkey applications: Development of technology for combustion of biomass/biomass-coal mixtures in a circulating fluidized bed. Project duration:

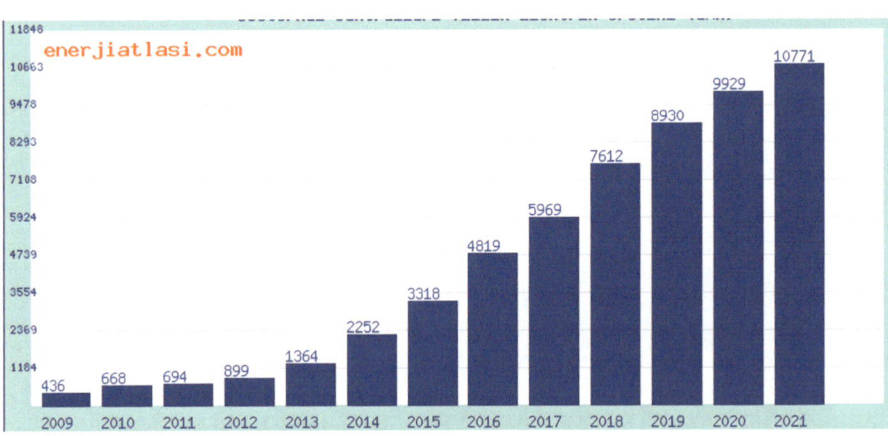

Figure 6.
Annual electricity production from geothermal power plants (GWh) [18].

3 years (September 2008–September 2011). Partners: TUBITAK, METU, GAMA Energy; TUBITAK KAMAG Program funds them. The project manager is from the METU Environmental Engineering Department. Institutions implementing the project are TUBITAK Marmara Research Center (MAM) and GAMA Energy. Project Objective: To develop "Cycled Fluidized Bed Combustion Technology" and create energy to take advantage of our biomass and lignite coals in the short and medium term to help spread cogeneration systems by co-producing heat and electricity. Production-knowledge package (Know-how); in this regard [21]. Project Outcomes: Deforestation and waste wood will produce pellets according to NTA8080 and FSC standards, and an export chain will be established. A regional supply chain will be created, and a promotional campaign for using wood pellets will be organized. The installed pellet production capacity is targeted at 6000 tons per month [22].

3.3 Projects contributing to the development of Turkey's renewable energy capacity

Some of the projects that Turkey is involved in to develop its renewable energy capacity are as follows [23]:

a. *National Wind Energy Systems Development and Prototype Turbine Manufacturing Project*: The project under TÜBİTAK's ruling process is to develop and manufacture prototypes of 2.5 MW industrial-scale wind turbines with original and indigenous technology.

b. *National Solar Power Plant Development Project*: The project aims to manufacture components required for solar-based electricity generation using indigenous technology. Within the framework of the UNDP/GEF-supported project titled.

c. *"Illumination of the Van Cat statue on the Van-Bitlis main highway with solar panels - Sustainable Energy Education in Turkey."* In July 1999, solar panels were installed to illuminate the Van Cat statue on the Van-Bitlis highway in the city of Van at night. A lighting system that works through panels has been built.

d. *LED and OLED Production Technologies Development Project*: The purpose of the project is to develop LED-OLED-based technologies used in lighting and display systems in the country and to ensure local production of prototypes using these technologies.

e. *National Hydroelectric Power Plant System Development Project*: Turbines, generators, etc. Work to be done in the project aimed at the internal development of power plant components. Elektrik Üretim A.Ş. It includes the restoration of all components of the Kepaz-1 Power Station (turbine, generator, turbine speed controller, generator warning system, and SCADA system) belonging to the General Administration.

f. *Project of LED lighting and Automatic Meter Reader System in street lighting*: a total of 218 units of conventional devices were replaced by LED within the project of energy-saving technologies by YEGM.

g. *"Rules and Principles for the Use of LED Illuminators in General Lighting"* was developed and entered into force on 23.01.2015.

h. *Green certificates*: YEGM reviewed studies conducted in the EU and other countries around the world regarding the development of legislation related to green certificates.

i. *Preparation and implementation of pilot projects based on hybrid systems*: It is planned to take maximum advantage of RES potential, ensure seamless integration into the electric power system, and increase the amount of energy to be obtained from existing and planned energy objects.

j. *Yahyali Pumped Hydroponic Supported Wind Energy Project (Hybrid)*: In 2013, the project, consisting of a wind farm and a pumped storage hydropower plant, aimed to maximize the potential of wind energy. Wind measurement data are recorded in the Wind Energy Monitoring and Forecasting Center system at the wind observation station, which provides wind data on-site.

k. *Construction and Operation Supervision Research Project*: The project aims to increase the production of electricity from wind energy sources and to ensure the integration of wind power plants into the Turkish Electricity System. Within the framework of Law No. 3096, YEGM provides technical consulting services by checking the compliance of private sector facilities with the approved projects.

l. *Renewable Energy Integration Project*: On May 9, 2014, a significant amount of funding was allocated for the project by the Board of Executive Directors of the World Bank. The project is helping to build submarine cable lines. Through these lines, smart grid investments will be made, and wind energy will be connected to all regions.

3.4 Incentive policies and practices for renewable energy

Promotion practices in Turkey have a structure that complements or supports development programs. When looking at the development programs implemented in Turkey since the establishment of the Republic until today, the elimination of regional disparities and efficient use of available resources are included in almost every program. Promotion measures have been implemented in Turkey in different periods, covering different practices and generally revised by different laws [24].

The Electricity Market Law No. 4628 was passed on February 20, 2001. Purpose of Electricity Market Law No. 4628: It is to ensure sufficient, quality, and uninterrupted electricity production to meet the needs of Turkey. It does this while keeping expenditures to a minimum and avoiding environmental damage. The objective is to create a financially robust, balanced, and accessible electricity market and independent oversight and regulation of this market.

Law No. 5346 aims to develop and grow power production from renewable energy sources such as wind, hydraulic, hydrogeological, geothermal, solar, and biomass (including landfill gas). Increase electricity production from renewable energy sources, ensure the most appropriate inclusion of these resources in the economy, and develop and manufacture the necessary tools and equipment for producing electricity from renewable energy—household appliances. Fees for renewable energy production

Types of facilities producing production based on renewable energies	Application price (USD/kWh)
a. Hydroelectric power generation facilities	7.3
b. Energy production facilities with wind	7.3
c. Energy production facilities with geothermal resources	10.5
d. Energy production facilities with biomass	13.3
e. Solar energy production facilities	13.3

Table 1.
Fixed price guarantees applied to renewable energy resources in Turkey [25].

facilities will be put into operation after the Council of Ministers determines 31.12.2015, provided they do not exceed the prices in **Table 1**.

The new Electricity Market Law No. 6446 introduced innovations and incentives, supplementing the shortcomings of Electricity Law No. 4628. License exemptions and tax incentives are granted to renewable energy facilities. Small generating units up to 5 MW do not require a license; municipal facilities and private surplus purchases are encouraged. Tax incentives were introduced for privatization operations and contracts. In addition, power generation enterprises are given a 50% discount and stamp duty exemption for 5 years.

The Energy Efficiency Law No. 5627 aims to lessen the impact of energy on the country's economy, which raises the effectiveness of energy sources and produced energy, and safeguards the environment. This law covers the procedures and principles to be applied in energy production enterprises and distribution networks, industrial and industrial enterprises, transportation organizations, residences, and buildings during energy production, transmission, and consumption, increasing energy efficiency and awareness in all segments. Energy efficiency projects are supported; the costs of the projects can be paid by 20 percent, and the president's decision can increase this percentage. If industrial enterprises undertake to reduce energy consumption by 10, 20% of their energy costs will be reimbursed. Waste treatment and renewable energy facilities are within the scope of incentives for enterprises that sign voluntary contracts. Enterprises with 500–1000 TOE energy consumption can also benefit from this support.

According to Law No. 5686 in force on geothermal resources, the principles of finding natural mineral or geothermal water resources in nature, enriching and protecting resources, have all exploitation rights over these resources or transferring these rights, as well as the principles of these enterprises operating within the framework of specific economic criteria without violating the conditions of natural harmony.

The Energy Sectors R&D Projects Support Program (ENAR) aims to support domestic energy production and technological development in Turkey through the policies set by the Ministry of Energy. The regulation framework encourages the continuous development of energy equipment and production methods, and a robust infrastructure is created by increasing employment and knowledge in this field [23]. Projects to be supported: It covers goals such as sustainable energy technologies, systems that reduce environmental damage, electricity transmission, and grid modernization. The ministry pays 80% of the project budget and 20% of the project owner. There are also grants from private companies and international programs [26]. Investment and support loans cannot be applied to ENAR initiatives [23].

Up to 30% of financial support is provided for industrial enterprises whose value does not exceed 1 million TL for efficiency-enhancing projects (EAP) [23]. A total of 15,922 TOE of energy and 13.8 million TL were saved, thanks to 28 VAP projects implemented between 2009 and 2015 [23]. In addition, investments with an annual energy consumption of at least 500 TOE and 20% energy savings per product can benefit from Region 5 Incentives [26].

3.4.1 Non-tax incentives for the use of renewable energy resources

The first renewable energy power plants established in Turkey were implemented within the framework of Law No. 18610 dated 04.12.1984 on Entrusting Electricity Production, Transmission, Distribution, and Trade to organizations other than the Turkish Electricity Authority.

Feed-in tariff incentives (addition of local contributions and fixed price guarantee): Feed-in tariff incentives have been developed to support renewable energy production by increasing the use of local equipment [27]. The addition of local equipment is encouraged for 5 years as an additional payment to the fixed price guarantee [31]. The price guarantee per kWh, updated on 07.01.2021, is valid for power plants that will be operated until 2025. The highest incentive was given to geothermal and biomass resources (54 kurus), and the lowest was to solar and wind energy (32 kurus). Incentive prices are updated four times yearly based on currency and inflation rates and expressed in pence. The new prices are reduced by 25 to 67% compared to the old YEKDEM incentives and are predicted to harm renewable energy investments [28].

Exemption from license fee: In Turkey, the License-Free Production Right is an important mechanism used to promote renewable energy investments. This right offers investors the opportunity to produce energy without paying a license fee and provides a license fee exemption for the first 8 years, after which a 10% payment is required [29–31]. This promotion contributes to reducing bureaucratic processes, increasing competition, and ensuring the security of the energy supply [27].

Incentive by tender method: The tender method was launched in 2016 with the YEKA program and aimed to promote the use of local equipment. The first tender was held in 2017 for a 500 MW solar power plant in Konya-Karapınar, and the second was for wind power plants. Tender conditions include using local equipment, hiring local engineers, and conducting R&D [31].

3.4.2 Tax-based incentives for the use of renewable energy

In order to popularize the use of RE investments, incentives such as excise duty exemption and property tax exemption applied in many countries of the world are not applied in Turkey [30].

Exemption from corporate tax and VAT: According to the temporary Article 3 of Law No. 6446, the profit arising from the transfer, division, partial division, and merger operations, which will be carried out within the framework of the privatization of electricity distribution companies and electricity generation companies until the end of 2023, is exempt from corporate tax [27]. In addition, deliveries and services to be carried out under the temporary Article 3 of the same law are also exempt from VAT.

Exemption from stamp duty: Exemption from stamp duty was the only tax incentive applied to RE investments in Turkey until 2012 [32]. Transactions made during the investment period of RE production enterprises are exempt from duty, and issued papers are exempt from stamp duty [27]. In addition, according to Law No. 6446, papers issued for transactions in organized wholesale electricity markets are exempt from stamp duty [31]. Under Decision No. 2012/3305, RE investments are supported by general, regional, large-scale, and strategic investment incentives [33]. General incentives include exemption from VAT and customs duty for all regions; regional incentives offer advantages such as tax reductions, insurance premium support, and allocation of investment location [34]. Large-scale investments are identical to regional incentives except for interest support and are not directly applicable to RE technologies. Strategic investments target products highly dependent on imports and require an investment of at least 50 million TL. The production of electricity from energy sources other than natural gas can be considered a strategic investment [33].

3.5 Turkey's National Renewable Energy Action Plan (2013–2023)

The National Renewable Energy Action Plan (REAP) covers 2011–2020. It is submitted to the European Commission to comply with the mandatory targets of the Renewable Energy Directive No. 2009/28/EC in the Energy section, which is one of the most critical chapters of Turkey as a candidate country for the European Union. To this end, due to the abundance of renewable energy sources such as hydro, wind, solar, biomass, and geothermal, the government has created a portfolio that can meet 30% of the total electricity demand from renewable energy in 2023 and 20% per capita energy, that is, energy efficiency also targeted [14].

National Renewable Energy Action Plan in Turkey [35], national renewable energy strategy, expected energy consumption for 2013–2023, renewable energy targets and development, and measures and steps to achieve the targets are discussed. Under the plan, electricity consumption will increase from 246 TWh in 2013 to 424 TWh in 2023, an increase of 72 percent, creating 61,000 MW of renewable energy installed capacity capable of producing about 159 TWh of electricity. According to the plan, 2013 data and 2023 targets for electricity generation and installed capacity are presented in **Table 2** below.

According to the plan, 34,000 MW for hydroelectric power, 20,000 MW for wind power, 1000 MW for geothermal power, 5000 MW for solar power, that is, photovoltaic and concentrated solar power, and 1000 MW for biomass. In addition, it aims to reduce CO_2 emissions to generate at least 30 percent of electricity from renewable sources in 2023, develop new industrial activities with the wide application of new renewable energy infrastructure, and show positive economic effects on the infrastructure, construction, and operation stages [35].

Renewable energy technology	Installed power capacity (MW)		Electricity production (GWh)	
	2013	2023	2013	2023
Hydroelectric	22.289	34.000	59.420	91.800
Wind	2759	20.000	7.558	50.000
Geothermal	310	1.000	1.364	5.100
Sun	0	5.000	0	8.000
Biomass	224	1.000	1.171	4.533

Table 2.
2013 data and 2023 forecasts of electricity production and installed power capacity according to Türkiye's National Renewable Energy Action Plan [35].

3.6 Research methodology, analysis, and evaluation

At this research stage, a questionnaire was conducted to evaluate society's general opinion and awareness of renewable energy resources and environmental sustainability. The questionnaire is aimed at the general public living in Turkey and professionals working in the energy sector. The data obtained from this survey will provide valuable information about adopting renewable energy sources and what needs to be done for environmental sustainability. The online survey consisted of 30 questions (28 multiple-choice and two open-ended). The survey was conducted digitally using Google Forms.

One hundred and four participants across Turkey participated in the survey online. The questionnaire results reveal society's view of renewable energy, awareness in this field, and social support for renewable energy projects. The findings also play an essential role in identifying barriers to renewable energy projects and strategies to increase social acceptance of these projects.

First of all, when we look at the demographic profile of the participants, we see that the survey consists mainly of young and middle-aged groups. When we look at the gender distribution, the number of male participants is higher than that of female participants.

According to the results of the questions regarding the level of knowledge about renewable energy, most participants reported having good (48%) or high (27.9%) knowledge about renewable energy resources. However, 24.1% of the group have an average or low level of knowledge. This situation shows that although the level of knowledge is generally high, a group also needs information. This highlights the importance of awareness and awareness programs.

Among the sources of information, the internet and social media are the most preferred channels. It reveals how effective digital platforms are in disseminating renewable energy and sustainability information. It is seen that traditional media sources are less preferred (**Figures** 7 and **8**).

4. Results

The survey results in this second part assess participants' attitudes and beliefs toward renewable energy. These data help us understand how society perceives renewable energy, the environmental and economic impacts of these resources, and what technological developments it cares about. According to the results of the questions, 89.4% of the

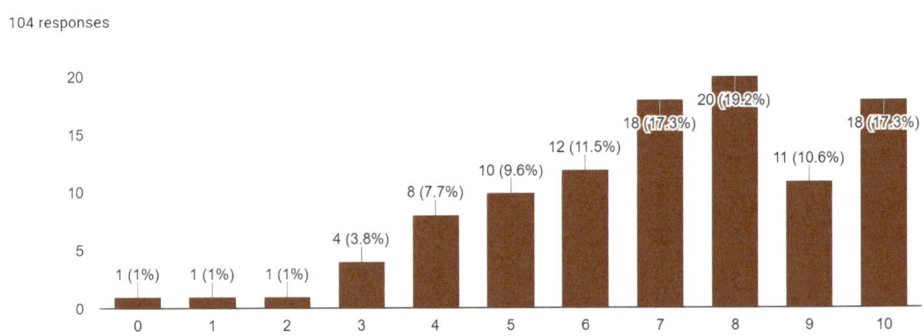

Figure 7.
Level of knowledge about renewable energy sources (Own elaboration). How would you rate your knowledge level about renewable energy sources?

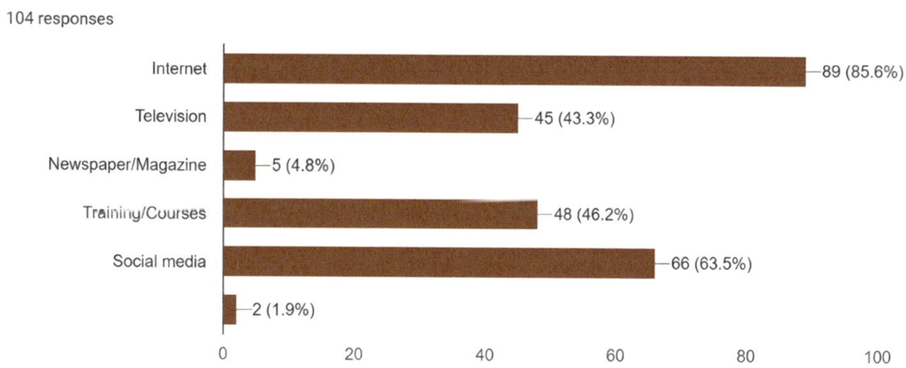

Figure 8.
Resources to learn about renewable energy (Own elaboration). What sources do you use to learn about renewable energy sources?

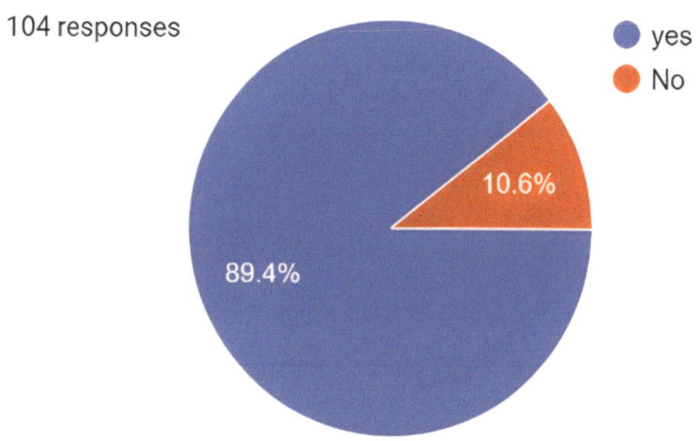

Figure 9.
The rate of support for renewable energy use (Own elaboration). Do you support the use of renewable energy?

participants supported using renewable energy (**Figure 9**). This high indicator shows that most of society has a positive attitude toward renewable energy.

Participants' perspectives on the environmental impact of renewable energy sources: 76% of participants rate the environmental impact of renewable energy sources as positive or very positive. This shows that renewable energy is widely accepted as an environmentally friendly alternative. However, 24% have a more cautious or negative assessment of this issue, suggesting that more information on environmental impacts should be provided (**Figure 10**).

Looking at the results of the question Contribution of Renewable Energy Projects to the Local Economy, it is clear that 74.9% of participants believe that renewable energy projects can contribute to the local economy. This shows a widespread perception that local economies can be strengthened through sustainable energy investments. About 25% expressed moderate to low beliefs, indicating that economic contributions are still uncertain for some (**Figure 11**).

4.1 Sustainable development and renewable energy

The survey results in this section allow for the finding of how participants perceive sustainable development and renewable energy and the main challenges in this area.

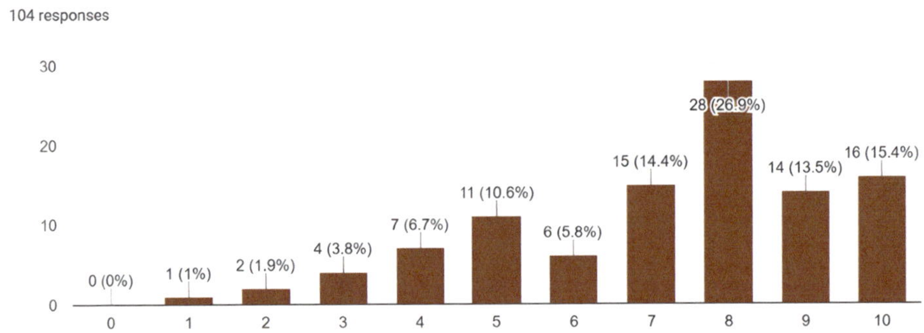

Figure 10.
Environmental impact assessment of renewable energy sources (Own elaboration). How do you evaluate the environmental impacts of renewable energy sources?

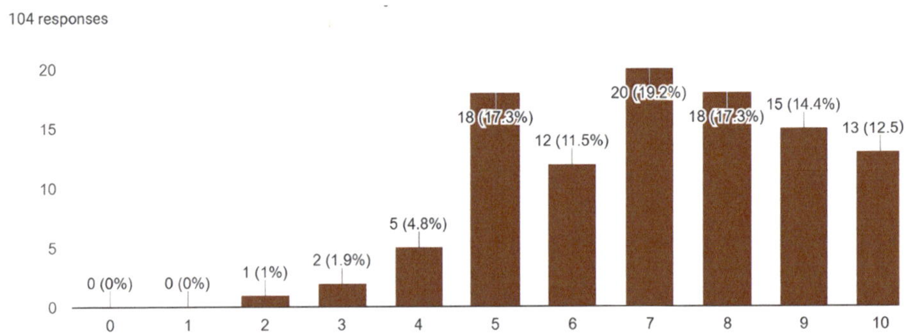

Figure 11.
Level of belief that renewable energy projects can contribute to the local economy (Own elaboration). Do you believe that renewable energy projects can contribute to the local economy?

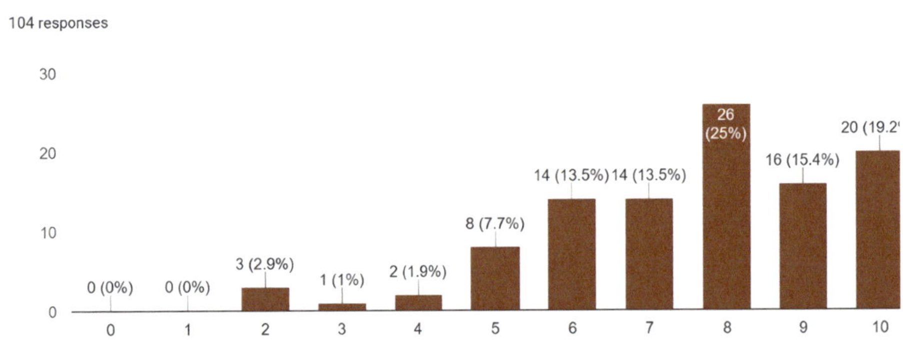

Figure 12.
Level of belief that the use of renewable energy can increase environmental sustainability (Own elaboration). Do you believe that the use of renewable energy can increase environmental sustainability?

According to the results of one of the questions, the majority (86.6%) strongly believe that renewable energy can increase environmental sustainability. In fact, 34.6 percent are entirely sure of this issue. However, 13.5% of the group is either undecided or has a negative opinion. This shows us that the belief in renewable energy is quite strong, but some individuals may still hesitate (**Figure 12**).

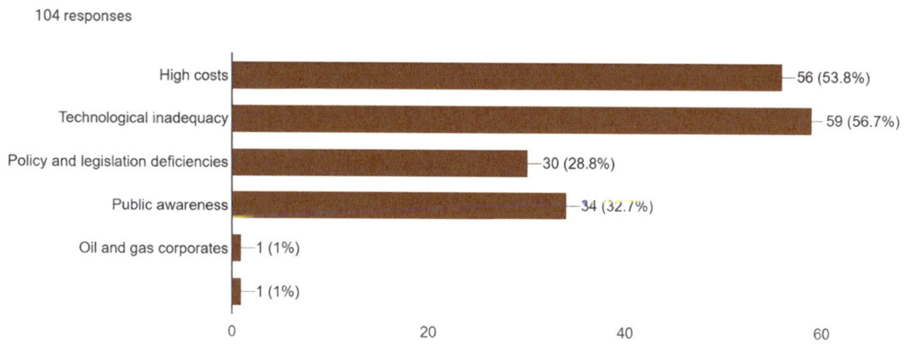

Figure 13.
The main challenges in implementing renewable energy projects (Own elaboration). What are the main challenges in implementing renewable energy projects?

This indicates that the technological infrastructure and funding required to implement the projects still need to be improved. In addition, factors such as lack of policy and legislation (28.8%) and insufficient awareness of the population (32.7%) are cited as other vital obstacles to these projects. These findings show that for the successful implementation of renewable energy projects, essential steps must be taken at the technological, political, and social levels (**Figure 13**).

4.2 Renewable energy and society

The survey results in this section assess the social acceptance of renewable energy projects and public attitudes toward providing financial support to these projects. It explains how participants approach renewable energy projects and what steps must be taken to make them widespread. Also, 42.3% of respondents think economic incentives are the most effective step to increase social acceptance of renewable energy projects. This finding suggests that providing a tangible economic benefit for the community to be more supportive of such projects is essential. In addition, 27.9% believe that public awareness campaigns will play an important role (**Figure 14**). This highlights the need for better public awareness of the benefits of renewable energy.

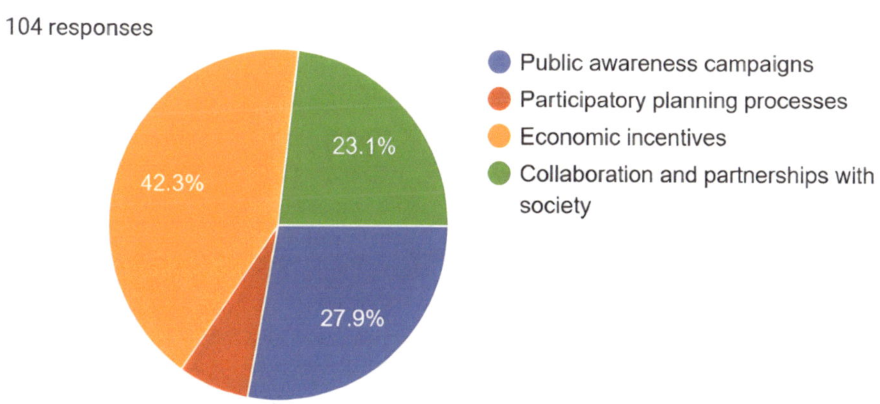

Figure 14.
Steps to be taken to increase social acceptance (Own elaboration). What steps should be taken to increase social acceptance of renewable energy projects?

According to the findings of Social Attitudes on Financial Support, the proportion of participants willing to provide financial support to renewable energy projects at the call of the government is 71.2%, which indicates that society as a whole is inclined to support such projects. However, 28.8% are hesitant on this issue (**Figure 15**). As a reason, the majority of those who do not want to provide financial support (46.7%) believe that the financing of such projects should be paid by the state (**Figure 16**). This reveals a strong belief that the state should be more active in financing renewable energy projects. Other reasons include the economic situation (17.8%) and the idea that current taxes are sufficient (15.6%).

The graph shows the percentage distribution of the amounts participants are willing to contribute to renewable energy projects. Most participants (33.7%) chose to contribute €20. This is followed by those willing to contribute €10 at 25.3% and €100 at 9.6%. The proportion of those who prefer to pay 30 euros is 12.04% (**Figure 17**). This trend shows that projects should be affordable and accessible regarding social support.

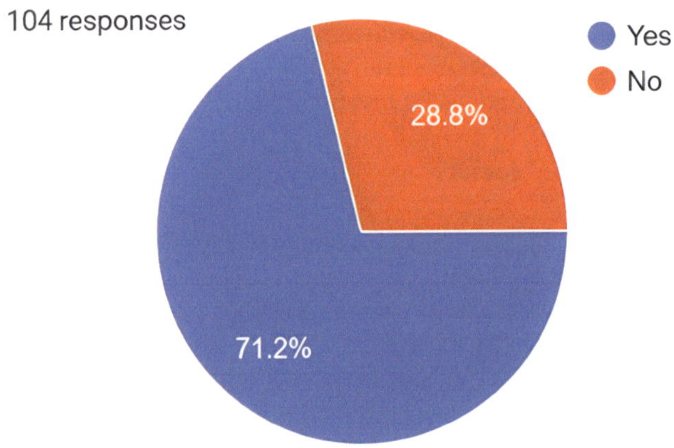

Figure 15.
Willingness to contribute financially to renewable energy projects (Own elaboration). If the government asks you to contribute financially to renewable energy projects, would you support these initiatives?

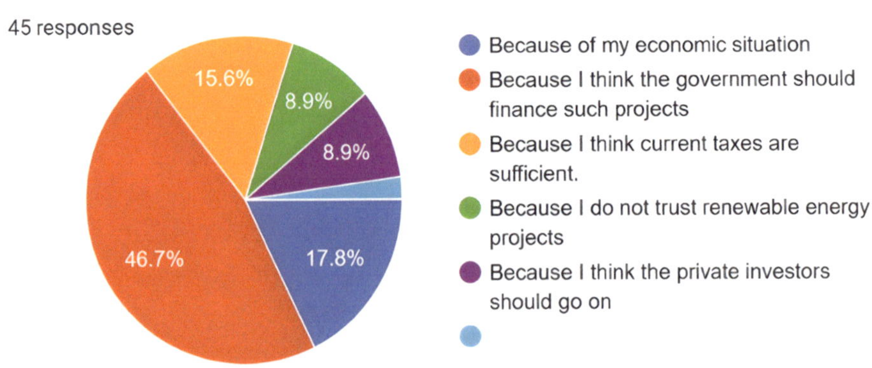

Figure 16.
Reasons for not contributing financially (Own elaboration). If you answered no, please select the reason.

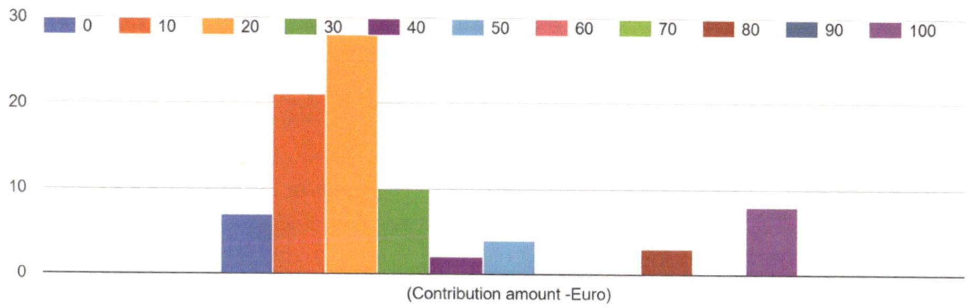

Figure 17.
Amounts of financial contributions (Euros) (Own elaboration). If you answered yes, how much would you like to contribute (one-time payment)?

5. Discussion

The findings of this survey show that the majority of participants in Turkey have a positive attitude toward renewable energy sources and believe that these sources are important for sustainable development and environmental sustainability. Most participants believe that using renewable energy can increase environmental sustainability and that these projects will contribute to the local economy. However, it was emphasized that there are also difficulties in implementing renewable energy projects, such as high costs and technological deficiencies.

In addition, the importance of education programs and public awareness campaigns to increase public awareness is emphasized. Although most participants expressed their willingness to contribute financially to renewable energy projects, they expressed that the government and the private sector should take more responsibility.

Although public support for renewable energy is generally strong, the successful implementation of these projects requires economic incentives, social cooperation, and investment in innovative technologies. These results emphasize the importance of strategic planning for integrating renewable energy and environmental sustainability.

Turkey and other leading countries in energy and economy are participating in the energy transformation. The National Energy and Mining Policy is the locomotive and acts as a general framework in determining the steps to be taken.

Turkey implements these decisions through legal regulations, international agreements, energy conservation directives, investment incentives, taxes, and other public policy methods. For example, the construction and operation inspection research project will provide the integration of wind energy power plants into the Turkish electricity system. With the LED and OLED Production Technologies Development Project, LED- and OLED-based technologies used in lighting and display systems will be developed in the country, and local prototypes will be produced using these technologies. Completing many notable projects like this and making more calls for projects will ensure that Turkey becomes one of the countries that make a name for itself.

Support for including more renewable energy sources in the energy portfolio and increasing localization in renewable energy technologies has been increased, and positive results have been observed. Today, Turkey has become a country that supplies more than 50% of its electricity demand from renewable energy sources.

In Turkey, the state should be able to act independently and make decisions freely in the energy field. This is extremely necessary to close the current deficit. In order to create a tax policy that supports renewable energy and energy efficiency, tax incentives should be applied to companies and individuals investing in RES, and the VAT rate for energy produced from RES should be reduced. In transport, a reduced VAT rate should be applied to rail and sea transport services that consume less energy, the tax burden on urban public transport services should be reduced, public transport services should be expanded, and private sector investments should be made in this area.

It is seen that Turkey is not indifferent to the negative impact of energy consumption on the environment; while moving to clean policies in the country, it is also trying to find a solution to the problem of climate change caused by global warming, with the international community contextually. As a result of Turkey's efforts, greenhouse gas emissions are expected to decrease by about 20% compared to normal projections by 2030.

In addition, the questionnaire conducted as part of this research assessed the public perspective of renewable energy, awareness of these energy sources' environmental and economic impacts, and social acceptance of renewable energy projects. The findings show that most societies support renewable energy projects and are aware of the positive effects of these projects on the environment. These findings are consistent with Turkey's energy policy analysis and point to the steps needed for renewable energy projects to gain broader support within the country. The increasing share of renewable energy sources in Turkey's energy portfolio and the support of society in this field will play an essential role in achieving the country's energy independence and sustainable development goals.

Furthermore, with solid targets set for 2030, Turkey's efforts to make significant progress in sustainable energy policies can be seen. Nevertheless, there are indeed many ways to go. It is necessary to put forward a sustainable long-term strategy in Turkey's energy field. Finally, the policies defined today in the energy field for Turkey, which is in a critical period, that will determine the country's place among the world's countries for the years 2030 and 2050.

6. Recommendations

These recommendations include strengthening energy policy, supporting technological development, education, and awareness activities, and improving financial and investment opportunities. For Strengthening Policy and Legal Regulations, more attractive legal arrangements should be made to promote the use of renewable energy. For example, the license conditions of the EMRA (Energy Market Regulatory Authority) should be more attractive.

So, to support investments, tax incentives can be increased, and investors can be exempted from taxes for a certain period. Also, by targeting technological development and R&D, more resources should be allocated to R&D to increase renewable energy potential by 2030. The R&D budget can be significantly increased.

The following suggestions can be made for education and awareness: The Ministry of National Education and NGOs should increase social awareness of renewable energy and sustainable development. Large-scale campaigns can be organized every year. Renewable energy engineering programs in universities can be increased to produce more graduates each year. Local governments should participate in renewable energy projects and provide resources to municipalities. Local economies should be supported by developing regional renewable energy solutions. For example, the potential of solar energy in Southeast Anatolia can be increased by 50%.

For example, solar-powered bicycles and vehicles should be promoted to encourage local people. Municipalities can subsidize the use of these vehicles or offer affordable rental options. Home solar systems should be widespread, and incentives and subsidies should be provided for installation costs. Awareness campaigns should be organized, and the importance of using renewable energy should be explained to the public.

7. Conclusions

Turkey acknowledges the environmental challenges posed by energy consumption and is actively transitioning toward cleaner policies. In alignment with the international community, the country is also seeking solutions to mitigate climate change driven by global warming. As a result of these efforts, Turkey aims to reduce greenhouse gas emissions by approximately 20% compared to projected levels by 2030.

In addition, the questionnaire conducted as part of this research assessed the public perspective of renewable energy, awareness of these energy sources' environmental and economic impacts, and social acceptance of renewable energy projects. The findings show that most societies support renewable energy projects and are aware of the positive effects of these projects on the environment. These findings are consistent with Turkey's energy policy analysis and point to the steps needed for renewable energy projects to gain broader support within the country. The increasing share of renewable energy sources in Turkey's energy portfolio and the support of society in this field will play an essential role in achieving the country's energy independence and sustainable development goals.

Furthermore, with solid targets set for 2030, Turkey's efforts to make significant progress in sustainable energy policies can be seen. Nevertheless, there are indeed many ways to go. It is necessary to put forward a sustainable long-term strategy in Turkey's energy field. Finally, the policies defined today in the energy field for Turkey, which is in a critical period, will determine the country's place among the world's countries for the years 2030 and 2050.

Author details

Ziyafat Guluzade[1], Anna Montini[1] and Rui Alexandre Castanho[2,3]*

1 Department of Economics, University of Bologna, Italy

2 WSB University, Dabrowa Górmicza, Poland

3 VALORIZA—Research Centre for Endogenous Resource Valorization, Portalegre Polytechnic University, Campus Politécnico, Portalegre, Portugal

*Address all correspondence to: acastanho@wsb.edu.pl

References

[1] Altuntaşoğlu ZT. Türkiye'de Rüzgâr Enerjisi, Mevcut Durum, Sorunlar. Makine Dergisi. 2011;**52**(617):56-63

[2] Dişişleri, Bakanliği TC. Sürdürülebilir Kalkınma. In: Çevre, İklim Değişikliği ve Suya Dair Sürdürülebilir Kalkınma Hedefleri / T.C. Dışişleri Bakanlığı (mfa.gov.tr). Ankara, Turkey: Turkey Government; 2020

[3] UNDP. The 17 Goals. NY, USA: United Nations Development Programme; 2020. Available from: https://sdgs.un.org/goals

[4] Murat G. Sürdürülebilir Kalkınma 2030 Gündemi Bağlamında Çalışma Hayatında Cinsiyete Dayalı Ayrımcılık. Karadeniz Teknik Üniversitesi Sosyal Bilimler Enstitüsü Sosyal Bilimler Dergisi. 2017;**7**(13):7-36

[5] Kerk GV, Manuel AR. A comprehensive index for a sustainable society: The SSI — The sustainable society index. Ecological Economics. 2008;**66**(3):228-242

[6] Isaac M, Vuuren DP. Modeling global residential sector energy demand for heating and air conditioning in the context of climate change. Energy Policy. 2009;**37**(2):507-521

[7] Stanford A. A vision of a sustainable energy future. Renewable Energy. 1997;**10**(2):417-422

[8] Mongo M, Belaid F, Ramdani B. The effects of environmental innovations on CO_2 emissions: Empirical evidence from Europe. Environmental Science & Policy. 2021;**118**:1-9

[9] Raihan A, Muhtasim DA, Farhana S, Pavel MI, Faruk O, Rahman M, et al. Nexus between carbon emissions, economic growth, renewable energy use, urbanization, industrialization, technological innovation, and Forest area towards achieving environmental sustainability in Bangladesh. Energy and Climate Change. 2022;**3**:1-15

[10] Kaygusuz K, Kaygusuz A. Renewable energy and sustainable development in Turkey. Renewable Energy. 2002;**25**:431-453

[11] Kavcıoğlu Ş. Enerji Sektöründe Yatırım Projelerinin Değerlendirilmesi. İstanbul: Türkmen Kitapevi; 2015

[12] Cakarel E, House J. IPP investment in Turkey's electric power industry. In: Program on Energy and Sustainable Development, Working Paper. Vol. 2. Stanford, USA: Stanford University; 2004

[13] Kocaman AS. Pompaj depolamalı hibrid enerji sistemi optimizasyonuTürkiye için vaka analizi. Gazi Üniversitesi Mühendislik Mimarlık Fakültesi Dergisi. 2019;**34**(1):53-68

[14] ETKB. Stratejik Plan 2019-2023. Ankara, Turkey: Turkey Government; 2021. Available from: https://sp.enerji.gov.tr

[15] Bayraktar A. T.C. Enerji Ve Tabii Kaynaklar Bakanligi. 2025. Available from: https://enerji.gov.tr/

[16] Savrul K. Burcu. In: Enerji Ekonomisi: Türkiye'nin Enerji Sektörü ve Alternatif Enerji Kaynakları. Çanakkale: Dora Basım Dağıtım; 2016

[17] Statista. Empowering people with data. 2025. Available from: https://www.statista.com/

[18] Atlasi E. E-Bulten. Turkey: Enerji Atlası; 2025. Available from: https://www.enerjiatlasi.com

[19] Yilanci A, Dincer I, Ozturk HK. A review on solar-hydrogen/fuel cell hybrid energy systems for stationary applications. Progress in Energy and Combustion Science. 2009;**35**(3):231-244

[20] Yeşil M. TRA2 Bölgesindeki Yeşil Enerji Kaynakları Sektör Raporu. Ankara, Turkey: T.C. Serhat Kalkınma Ajansı; 2015. pp. 1-78

[21] Çağlar M. Dünya ve Türkiye'de Yenilenebilir Enerji Kaynakları. Elektrikişleri Etüt idaresi Genel Müdürlüğü. Ankara, Turkey. p. 20

[22] Belen İ. Orman Biyokütlesinden Enerji Üretimi. İzmir: Uluslararası Enerji Türk Zirvesi; 2010

[23] T. C. Enerji ve Tabii Kaynaklar Bakanlığı. Stratejik Plan: 2015-2019. 2014. Available from: http://sp.enerji.gov.tr/ETKB_2015_2019_Stratejik_Plani.pdf [Accessed: April 29, 2016]

[24] Emre E. Türkiye'de Teşvik Sistemleri ve Mevcut Sistemin Yapısına Yönelik Öneriler. Ankara, Turkey: T.C Başbakanlık Devlet Planlama Teşkilatı, İktisadi Sektörler ve Koordinasyon Genel Müdürlüğü, Ankara; 2011. pp. 35-55

[25] T.C. Resmi Gazete. 5346 Sayılı Yenilenebilir Enerji Kaynaklarının Elektrik Enerjisi Üretimi Amaçlı Kullanımına İlişkin Kanun. Ankara, Turkey: Turkey Government; 2005. 534. Available from: https://www.mevzuat.gov.tr/MevzuatMetin/1.5.5346.pdf

[26] Kurban E. Uygun Görülen Verimlilik Artırıcı Projeler! Tam Liste!. 2021. Available from: https://www.enerjiportali.com/2021-uygun-gorulen-verimlilikartirici-projeler-tam-liste/

[27] İbrahim D. Yenilenebilir Enerji Kaynakları, Muhasebe, Vergi Uygulamaları, Sektöre Sağlanan Teşvik ve Hibeler Ankara. Vol. 133. Ankara, Turkey: Gazi Kitabevi; 2018

[28] Dunya, 2021: Dunya. Dunia, Turkey; 2021. Available from: www.dunya.com

[29] Olcay Y, Hakan H. Yenilenebilir Enerjiye Yönelik Teşvikler ve Türkiye. Adnan Menderes Üniversitesi Sosyal Bilimler Enstitüsü Dergisi. 2015;**2**(2):74-97

[30] Eser Levent Yahya ve Polat Sedat. Elektrik Üretiminde Yenilenebilir Enerji Kaynaklarının Kullanımına Yönelik Teşvikler: Türkiye ve İskandinav Ülkeleri Uygulamaları. Gümüşhane Üniversitesi Sosyal Bilimler Elektronik Dergisi. 2015;**12**:201-225

[31] Berksoy Turgay ve Akbaş Akdoğan. Yenilenebilir Enerjide Kamu Politikaları ve Türkiye. Journal of Life Economics. 2018;**5**(3):19-42

[32] Sezer Y. Enerjide Yatırımcı ve Tüketici Fiyatlarının Vergi Boyutu. 14. Uluslararası Enerji Arenası. Ankara, Turkey: İstanbul; 2012

[33] Mustafa E, Abdullah T. Türkiye'de Teşvik Sisteminin Yapısı, Sorunları ve Etkinliği Üzerine Bir Politika Önerisi: Tek Bir Uygulamacı Kuruluş Sorunları Çözer mi? Atatürk Üniversitesi İktisadi ve İdari Bilimler Dergisi. 2018;**32**(3):725-744

[34] Asuman A, Şahin K. Türkiye'de Mali Teşvik Sistemi ve Yatırımlara Sağlanan Mali Teşviklerin Değerlendirilmesi. Adnan Menderes Üniversitesi Sosyal Bilimler Enstitüsü Dergisi. 2007;**4**(4):189-202

[35] T.C. ETKB. Türkiye Ulusal Yenilenebilir Enerji Eylem Planı. Ankara: Türkiye Cumhuriyeti Enerji ve Tabii Kaynaklar Bakanlığı; 2014

Chapter 8

Perspective Chapter: Challenges and Future Trends in Wireless Power Transfer for Biomedical Implants

Abdenasser Lamkaddem, Ahmed El Yousfi and Daniel Segovia Vargas

Abstract

Wireless power transfer (WPT) in biomedical implants has become a vital area of research and innovation, primarily because it eliminates the need for frequent and invasive battery replacements. Biomedical implants, such as pacemakers, neural stimulators, and drug delivery systems, require reliable and long-term power sources to ensure their functionality. Traditionally, these devices relied on batteries that necessitated surgical interventions for replacement once implanted. However, such procedures pose risks of infection, tissue damage, and patient discomfort. WPT offers a non-invasive solution to this challenge by enabling energy transmission through electromagnetic fields or other techniques. Various methods have been proposed and studied in the literature for transferring power wirelessly within the human body. Capacitive coupling, inductive coupling, and resonant inductive coupling are among the most common methods. Each type has its own advantages and limitations based on factors such as range and power efficiency. By advancing WPT technology, researchers are enhancing the reliability, safety, and convenience of biomedical implants, improving the quality of life for millions of patients worldwide.

Keywords: battery, biomedical implants, implantable antenna, rectifier, rectenna, wireless power transfer

1. Introduction

WPT in wireless biomedical implants gained a lot of attention because it is a promising solution to power the implanted device. Devices such as pacemakers, capsule endoscopy, intracranial pressure monitoring, insulin drug pumps, deep brain neurostimulators, and gastric stimulators may need WPT to overcome the battery replacement problems.

One of the key challenges of these devices is establishing communication between a sensor that monitors the patient's condition and an external control unit.

Traditionally, this communication has relied on wired devices, which can lead to issues such as infection, device displacement, and patient discomfort [1]. In contrast, using a wireless device minimizes these risks, making it a more suitable option for implantable applications. Other challenges arise due to the harsh environment of the human body, which includes factors such as fluctuating temperatures, constant motion, the presence of bodily fluids, and potential interference from electromagnetic fields. In addition, the impossibility of knowing the orientation of the implanted device from outside. These environmental conditions can significantly impact the performance, longevity, and safety of the devices, requiring careful consideration during their design and operation.

Supplying power to a Wireless Implantable Medical Device (WIMD) located deep within the body raises questions about the reliability and efficacy of the WPT system. To evaluate the effectiveness of the power transfer, the main performance metric used is Power Transfer Efficiency (PTE). PTE is defined as the ratio of the power successfully delivered to the implant's load compared to the total power transmitted from the source [2].

$$PTE = \frac{P_r}{P_t} = \frac{k_{12}^2 Q_1 Q_2}{1 + k_{12}^2 Q_1 Q_2} \quad (1)$$

where P_r presents the received power and P_t denotes the transmitted power. The coupling factor k_{12} between the coils, along with the quality factors Q_1 and Q_2 of the Transmitter (TX) and Receiver (RX) components, respectively.

The received power can be determined using

$$P_{Rx} = P_T + G_T + G_r - L_P - 20\log_{10}\left(\frac{4\pi d_0}{\lambda_0}\right) - 10\gamma\log_{10}\left(\frac{d}{d_0}\right) - X_\sigma \quad (2)$$

where P_T is the input TXpower, G_T is the implantable antenna gain at the resonant frequency, G_r is the RX antenna gain, $L_P = 0.3$ is the polarization loss, $\gamma = 1.5$ is the path loss exponent, $X_\sigma = 2.2$ is the shadowing factor [3].

The RF to DC conversion efficiency η (%) is calculated by [4, 5].

$$\eta(\%) = \frac{P_{out}}{P_{in}} \times 100\% = \frac{I_{out} \times V_{out}}{P_{in}} \times 100\% \quad (3)$$

where P_{in} and P_{out} are the input power and the output power, respectively. V_{out}, I_{out} are the output voltage and current, respectively.

Different techniques exist for transferring energy wirelessly inside the human body. They are divided into two types. The first one is ultrasonic energy transfer (UFT), which uses acoustic waves to transmit energy, and the second one is the electromagnetic power transfer (EMPT), which is also divided into two methods. The far-field electromagnetic power transfer and the near-field power transfer. Each technique has its advantages and drawbacks that we will explore later in this chapter.

2. Types of WPT

As mentioned before, there are two major methods in EMPT.

2.1 Far-field electromagnetic power transfer

Far-field electromagnetic power transfer refers to the wireless transmission of electrical energy over considerable distances by means of electromagnetic waves [6–10]. Unlike near-field techniques that work over short ranges, far-field power transfer operates within the "far-field" region, where electromagnetic waves can propagate freely and efficiently over larger distances $R \geq 2D^2 / \lambda$.

where λ is the operating wavelength and D is the diameter of the antenna.

This method allows energy to be transmitted from a source to a remote receiver, typically ranging from a few meters to several kilometers, without the need for physical connections [11]. Far-field electromagnetic power transfer has promising applications in powering remote sensors, unmanned aerial vehicles, and other electronic devices located far from traditional power sources.

2.2 Near-field power transfer

Near-field electromagnetic power transfer is a method of wireless energy transmission over short distances by utilizing electromagnetic fields within the "near-field" region, where the transmitting and receiving devices are close enough for magnetic or electric fields to interact directly. Near-field power transfer operates effectively within ranges of a few millimeters to a few meters, enabling efficient energy transfer over these limited spans. This technology is commonly used in applications such as wireless charging for smartphones, medical implants, electric vehicles, and other electronic devices where proximity between the transmitter and receiver ensures minimal energy loss and high transfer efficiency [12]. It is divided into three methods: Capacitive coupling. Inductive power transfer and magnetic resonance coupling.

Capacitive coupling: This technique transfers energy using capacitive couplers between the receiver and the transmitter. It relies on the transmission of electrical energy through the electric field generated between two conductive plates [13]. It is commonly used in touchscreens, sensors, and other electronic devices. However, the drawback of this method is its requirement for proximity between the transmitter and receiver, as well as its low efficiency. A theoretical model was developed using two CU plates. This analysis allows for the calculation of the PTE [14]. It was concluded that the PTE for capacitive coupling depends on factors such as the plate area, the distance between them, tissue thickness, and the operating frequency.

Inductive Power Transfer (IPT) is a longstanding technique for wireless power transmission. This method operates on the principle that an alternating current passing through the TX coil creates a magnetic field, which in turn induces a voltage in the RX coil. IPT has been applied across various implantable medical devices, including those for blood glucose monitoring [15], neural stimulation [16], pacemakers [17], cochlear implants [18], and cortical implants [19].

A PTE of 0.002% was achieved by designing a system with three distinct layers: a superstrate layer, an implantable antenna in the second layer using a Planar Inverted-F Antenna (PIFA), and a rectifier as the third layer [20]. A compact two-stage voltage doubler rectifier measuring 5 × 10 mm^2 was developed with a quad-band

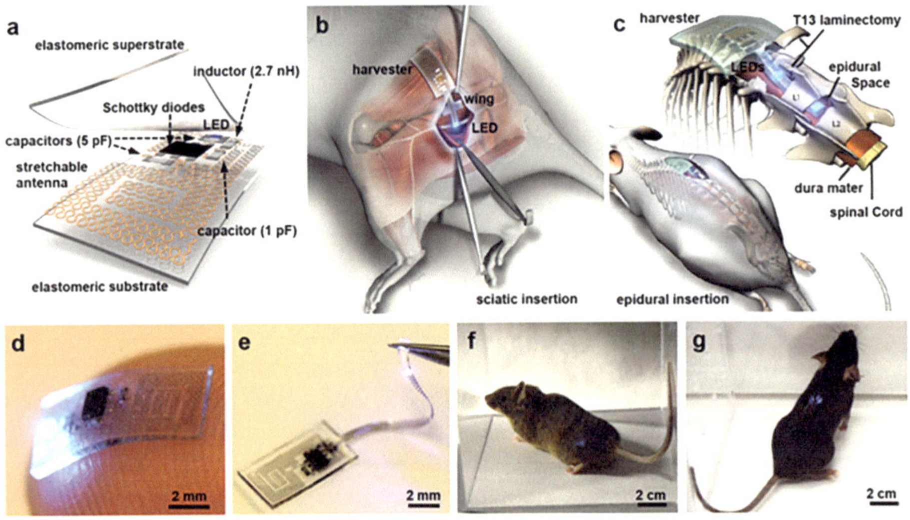

Figure 1.
A spinal cord implantable device on a rat [24].

implantable antenna, achieving a PTE of 0.67% [7]. To boost the PTE to an average of 26%, a closed-loop design was developed [21]. The design enables real-time control of a drug delivery system within a transfer range of 20 mm.

A PTE of 47.2% was reported using magnetic resonant coupling with spiral coil antennas on both the TX and RX ends [22]. In another study, Complementary Metal-Oxide Semiconductor (CMOS) technology was used at a carrier frequency of 13.56 MHz, resulting in a high PTE of 92.6%, though the design was relatively bulky [23].

A separate study implanted a $6 \times 3.8 \times 0.7\ mm^3$ device into the spinal cord of a rat, utilizing a stretchable antenna for the implantable component, as shown in **Figure 1** [24].

Magnetic resonance coupling, as its name suggests, uses magnetic resonance. The emitter coil and the receiving coil resonate at the same frequency [25]. When close enough, they can exchange power through the magnetic field generated between them. The advantages of this technique include power transfer over distances of up to a few meters and higher PTE compared to other methods. However, it has a limited transfer range and is sensitive to alignment. This method has been used in various applications, including medical implants, electric vehicles, and smartphones [26]. **Table 1** summarizes each technical detail, including its receiver and transmitter types, frequency range, transfer range, advantages, drawbacks, and applications.

3. Challenges in WPT

WPT for implantable devices offers significant advantages in improving the quality of life for patients. It helps reduce the need for battery replacements, which, in some cases, require highly complicated surgical procedures. However, WPT faces several challenges, which are fourfold:

> *Miniaturization: The device needs to be implanted inside the human body. It is required that the entire implanted device occupy no more than 2% of the volume of the targeted human organ [27].*

WPT types	**Far-field WPT**		**Near-field WPT**	
Methods		**Capacitive coupling**	**Inductive coupling**	**Magnetic resonance coupling**
Rx and Tx	Antennas	Metal plates	Coils	Resonators
Frequency range	GHz	MHz	KHz-MHz	KHz-MHz
Distance	m-Km	Mm	mm-cm	cm-m
Advantages	Resistant to misalignment Extended transfer range	Free from eddy current losses	Easily integrated	Resistant to misalignment High transfer and conversion efficiency
Drawbacks	Minimal transfer and conversion efficiency	High voltages	Limited transfer range Highly sensitive to misalignment	Sensitivity to Alignment Limited transfer range Relatively large coils
Applications	Low power wireless sensors Solar power system	Medical devices Railway vehicles	Smartphones Medical devices	Smartphones Medical devices Electrical vehicles
Examples	[6–11]	[13, 14]	[7, 15–24]	[25, 26]

Table 1.
Different WPT types.

Transfer range: Some human organs are located deep within the body, requiring WPT systems to operate efficiently over greater distances [28].

Orientation: The implanted device may rotate or shift inside the human body, making consistent energy transfer a challenging task [29].

Losses: The human body is a lossy medium, which significantly degrades the efficiency of energy transfer within it [30].

Biocompatibility and safety

Thermal considerations: Energy transfer can produce heat, which must be carefully managed to avoid damaging tissues. The system should adhere to Specific Absorption Rate (SAR) safety standards. IEEE C95.1–1999 standard (average SAR < 1.6W/kg over 1g) and IEEE C95.1–2005 standard (average SAR < 2 W/kg over 10g) to ensure surrounding tissues remain unharmed [31]. Material Compatibility: Materials used in the implant must be biocompatible and resistant to degradation in the human body over time.

Energy efficiency

The efficiency of power transfer is critical because energy loss not only reduces performance but also may cause additional heating of tissues. Designing systems with high PTE in lossy environments like the human body is a key challenge.

System reliability and longevity

Implantable devices must function reliably for extended periods without failure. This includes robustness against mechanical stress, corrosion, and long-term wear.

Dynamic human environment

Variations in body movements, changes in posture, and tissue composition can affect the alignment and coupling efficiency of WPT systems. The system must adapt to these variations dynamically.

Electromagnetic Interference (EMI)

External electromagnetic fields from other devices could interfere with the WPT system. Ensuring proper shielding and robustness against EMI is necessary to maintain reliable performance.

Power management and scalability

Managing the power delivered to different components of the implantable device while ensuring scalability for multi-functional systems, such as those combining sensing, actuation, and communication, is an ongoing challenge.

4. Trends in WPT

The meandered line technique has been widely used in implantable antennas due to its advantages in achieving a small size, as shown in **Figure 2** [32, 33]. Another technique to further miniaturize the antenna is the use of a defected ground plane [34]. The implantable antenna measures $0.03 \times 0.03\ \lambda^2$. The WPT system operating at 1900 MHz achieves a PTE of 0.25%.

Metasurfaces can enhance the PTE of a system by improving the coupling between the receiver and the transmitter. A circular patch antenna was investigated [35]. The WPT system, operating at 915 MHz, achieved a PTE of 0.37% with a transfer range of 7 mm. A metasurface on the skin was presented [36]. This metasurface, composed of rectangular and circular loops, improved the coupling between the receiver and transmitter by 12 dB, significantly enhancing the PTE of the WPT system. Recently, a flexible mu-negative metasurface (see **Figure 3**) was developed, which increased the PTE of the system by 17% [37].

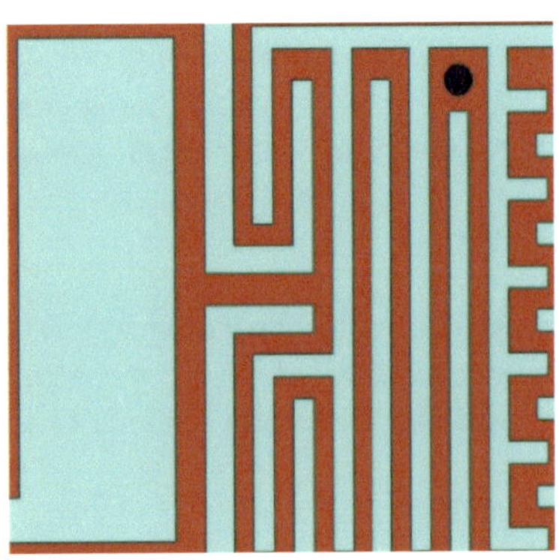

Figure 2.
Implantable antenna using meandered line technique for miniaturization [32].

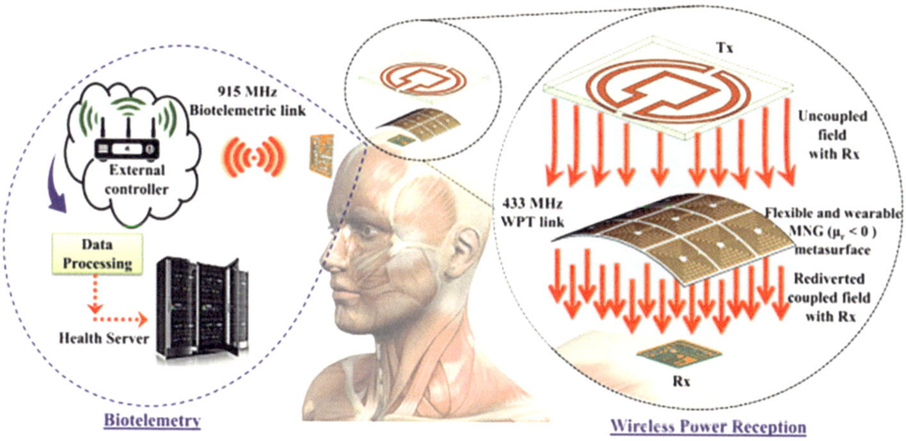

Figure 3.
Schematic of the WPT using the mu-negative metasurface [37].

The same topology was further studied, where the metasurface was designed with a multi-coil topology to increase the PTE to 2.32 and 2.6% at 6.78 MHz [38]. To avoid making the IMD (Implantable Medical Device) bulky and to reduce the risk of infections, a viable solution is to design a battery-free device. A battery-free implanted device with dimensions $16 \times 6.67\,mm^2$ was developed for WPT applications. This design features transparent encapsulation, a biocompatible substrate, and flexibility [39].

CMOS technology is a promising technique for WPT, as it enables compact designs. A fully implantable system with dimensions $4.7 \times 4.7\,mm^2$ for neural recording applications was reported [40]. A system-on-chip design was developed, combining radio-frequency components with other electronic elements in an area of $1.13 \times 0.3\,mm^2$ [41].

Musical acoustic vibration would be an effective way to transfer energy inside the human body [42]. This technique uses sound to excite a transponder. This reduces the misalignment sensitivity and increases the transfer range.

Multi input multi output (MIMO) rectenna technology offers the potential for high data rates in implantable devices. A cubic MIMO rectenna was developed for far-field WPT as shown in **Figure 4**. This design supports both right hand circular polarization (RHCP) and left hand circular polarization (LHCP), enabling energy harvesting for deep implantations and enhancing the rectifier efficiency to 70% [43].

3D coils provide an effective solution to enhance PTE, addressing issues related to orientation and misalignment. An omnidirectional WPT system was developed to generate a 3D magnetic field, achieving a PTE of 8.4% [44]. Similarly, a 3D multi-resonator WPT system (see **Figure 5**), where coils surround the human body to transfer energy over deep distances at 13.56 MHz. Using this method, the PTE can reach up to 20% throughout the entire human body, as reported [45].

In colonoscopy, the patient swallows a capsule through the mouth, which then travels through the stomach, intestine, and colon, sending images to an external device. To determine the origin of each image within the gastrointestinal tract, a tracking method is essential. The proposed method tracks the capsule's position by analyzing phase shifts in S-parameters across multiple channels. This solution could also help determine the orientation of the capsule inside the human body, improving the application of battery-free capsules and enabling efficient external power delivery [46].

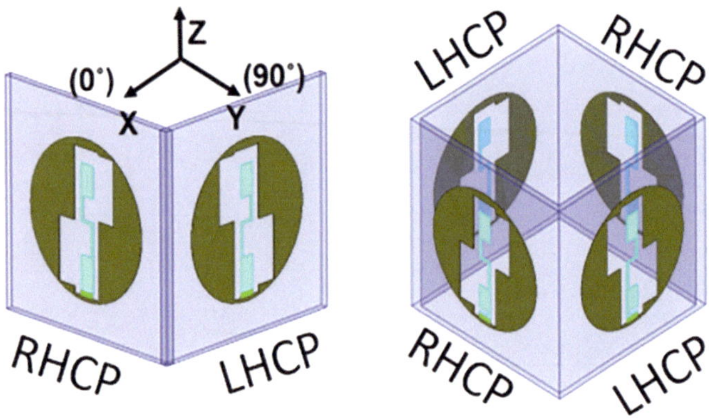

Figure 4.
The proposed 3D dual polarization implantable antenna [43].

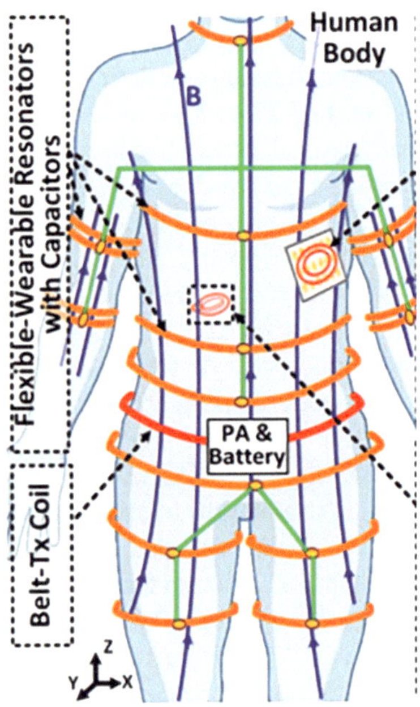

Figure 5.
3D multi-resonator WPT system [45].

Freq (MHz)	1470	1470	403	13.56/60	5
Types	Patch	Patch	Split Ring Loops	Coil	Coil
Tr. Distance (mm)	50	60	6	5	50
PTE (%)	0.67	1.07	57.9	0.24/0.36	40
Reference	[7]	[8]	[47]	[48]	[49]

Table 2.
State-of-the-art comparison.

Table 2 compares different state-of-the-art works. The distances vary between a few millimeters to several centimeters. However, the PTE differs significantly. In some cases, the authors achieve very limited efficiency, less than 1%, while in others, a high efficiency of around 59% is achieved using coils.

5. Conclusion

In conclusion, WPT plays a crucial role in advancing implantable devices by addressing challenges such as miniaturization, transfer range, efficiency, and safety. WPT technologies are categorized into far-field and near-field approaches, with near-field WPT further divided into inductive coupling, capacitive coupling, and magnetic resonance coupling techniques. Emerging trends in this field include the development of battery-free devices to minimize infection risks and device size, the use of CMOS technology for enhanced miniaturization and safety, and the integration of metasurfaces on the skin to improve coupling efficiency between transmitters and receivers. These advancements hold significant promise for the future of implantable medical devices.

Author details

Abdenasser Lamkaddem, Ahmed El Yousfi and Daniel Segovia Vargas*
Signal Theory and Communication Department, University Carlos III of Madrid, Madrid, Spain

*Address all correspondence to: dansevar@ing.uc3m.es

References

[1] Aggarwal RK, Connelly DT, Ray SG, Ball J, Charles RG. Early complications of permanent pacemaker implantation: No difference between dual and single chamber systems. British Heart Journal. 1995;**73**(6):571-575. DOI: 10.1136/hrt.73.6.571

[2] Lamkaddem A. Miniaturization of antennas and wireless power transfer system for biomedical implantable devices (Dissertation). Universidad Carlos III de Madrid; 2024. Available from: https://hdl.handle.net/10016/44301

[3] Jiang ZH, Gregory MD, Werner DH. Design and experimental investigation of a compact circularly polarized integrated filtering antenna for wearable biotelemetric devices. IEEE Transactions on Biomedical Circuits and Systems. 2016;**10**(2):328-338. DOI: 10.1109/TBCAS.2015.2438551

[4] Song C, López-Yela A, Huang Y, Segovia-Vargas D, Zhuang Y, Wang Y, et al. A novel quartz clock with integrated wireless energy harvesting and sensing functions. IEEE Transactions on Industrial Electronics. 2019;**66**(5):4042-4053. DOI: 10.1109/TIE.2018.2844848

[5] Wang X, Krois I, Ha-Van N, Mirmoosa MS, Jayathurathnage P, Hrabar S, et al. Time-varying systems to improve the efficiency of wireless power transfer. Physical Review Applied. 2024;**21**:054027

[6] Bhamra H, Tsai J-W, Huang Y-W, Yuan Q, Shah JV, Irazoqui P. A subcubic millimeter wireless implantable intraocular pressure monitor microsystem. IEEE Transactions on Biomedical Circuits and Systems. 2017;**11**(6):1204-1215. DOI: 10.1109/TBCAS.2017.2755596

[7] Basir A, Yoo H. Efficient wireless power transfer system with a miniaturized quad-band implantable antenna for deep-body multitasking implants. IEEE Transactions on Microwave Theory and Techniques. 2020;**68**(5):1943-1953. DOI: 10.1109/TMTT.2020.2965938

[8] Iqbal A, Al-Hasan M, Mabrouk IB, Basir A, Nedil M, Yoo H. Biotelemetry and wireless powering of biomedical implants using a rectifier integrated self-diplexing implantable antenna. IEEE Transactions on Microwave Theory and Techniques. 2021;**69**(7):3438-3451. DOI: 10.1109/TMTT.2021.3065560

[9] Zhou J, Zhang P, Han J, Li L, Huang Y. Metamaterials and metasurfaces for wireless power transfer and energy harvesting. Proceedings of the IEEE. 2022;**110**(1):31-55. DOI: 10.1109/JPROC.2021.3127493

[10] Shah SAA, Basir A, Lim Y-H, Yoo H. A novel efficient wirelessly powered biotelemetric endovascular aortic stent antenna system. IEEE Transactions on Antennas and Propagation. 2023;**71**(9):7132-7145. DOI: 10.1109/TAP.2023.3291419

[11] Xia M, Aissa S. On the efficiency of far-field wireless power transfer. IEEE Transactions on Signal Processing. 2015;**63**(11):2835-2847. DOI: 10.1109/TSP.2015.2417497

[12] Besnoff J, Abbasi M, Ricketts DS. High data-rate communication in near-field RFID and wireless power using higher order modulation. IEEE

Transactions on Microwave Theory and Techniques. 2016;**64**(2):401-413. DOI: 10.1109/TMTT.2016.2515586

[13] Park C et al. Separated circular capacitive coupler for reducing cross-coupling capacitance in drone wireless power transfer system. IEEE Transactions on Microwave Theory and Techniques. 2020;**68**(9):3978-3985. DOI: 10.1109/TMTT.2020.2989118

[14] Hossain ANMS, Mohseni P, Lavasani HM. Design and optimization of capacitive links for wireless power transfer to biomedical implants. IEEE Transactions on Biomedical Circuits and Systems. 2022;**16**(6):1299-1312. DOI: 10.1109/TBCAS.2022.3213000

[15] Lucisano JY, Routh TL, Lin JT, Gough DA. Glucose monitoring in individuals with diabetes using a long-term implanted sensor/telemetry system and model. IEEE Transactions on Biomedical Engineering. 2017;**64**(9):1982-1993. DOI: 10.1109/TBME.2016.2619333

[16] Lo Y-K, Kuan Y-C, Culaclii S, Kim B, Wang P-M, Chang C-W, et al. A fully integrated wireless SoC for motor function recovery after spinal cord injury. IEEE Transactions on Biomedical Circuits and Systems. 2017;**11**(3):497-509. DOI: 10.1109/TBCAS.2017.2679441

[17] Lee S-Y, Hsieh C-H, Yang C-M. Wireless front-end with power management for an implantable cardiac microstimulator. IEEE Transactions on Biomedical Circuits and Systems. 2012;**6**(1):28-38. DOI: 10.1109/TBCAS.2011.2162409

[18] Yip M, Jin R, Nakajima HH, Stankovic KM, Chandrakasan AP. A fully-implantable Cochlear implant SoC with piezoelectric middle-ear sensor and arbitrary waveform neural stimulation. IEEE Journal of Solid-State Circuits. 2015;**50**(1):214-229. DOI: 10.1109/JSSC.2014.2355822

[19] Silay KM, Dehollain C, Declercq M. A closed-loop remote powering link for wireless cortical implants. IEEE Sensors Journal. 2013;**13**(9):3226-3235. DOI: 10.1109/JSEN.2013.2259621

[20] Liu C, Guo Y-X, Sun H, Xiao S. Design and safety considerations of an implantable rectenna for far-field wireless power transfer. IEEE Transactions on Antennas and Propagation. 2014;**62**(11):5798-5806. DOI: 10.1109/TAP.2014.2352363

[21] Del Bono F et al. Design of a closed-loop wireless power transfer system for an implantable drug delivery device. IEEE Sensors Journal. 2024;**24**(6):7345-7354. DOI: 10.1109/JSEN.2023.3270521

[22] Sun G, Muneer B, Li Y, Zhu Q. Ultracompact implantable design with integrated wireless power transfer and RF transmission capabilities. IEEE Transactions on Biomedical Circuits and Systems. 2018;**12**(2):281-291. DOI: 10.1109/TBCAS.2017.2787649

[23] Li X, Tsui C-Y, Ki W-H. A 13.56 MHz wireless power transfer system with reconfigurable resonant regulating rectifier and wireless power control for implantable medical devices. IEEE Journal of Solid-State Circuits. 2015;**50**(4):978-989. DOI: 10.1109/JSSC.2014.2387832

[24] Park SI, Brenner DS, Shin G, Morgan CD, Copits BA, Chung HU, et al. Soft, stretchable, "fully implantable miniaturized optoelectronic systems for wireless optogenetics,". Nature Biotechnology. 2015;**33**(12):1280-1286

[25] Cannon BL, Hoburg JF, Stancil DD, Goldstein SC. Magnetic resonant

coupling As a potential means for wireless power transfer to multiple small receivers. IEEE Transactions on Power Electronics. 2009;**24**(7):1819-1825. DOI: 10.1109/TPEL.2009.2017195

[26] Li H, Li J, Wang K, Chen W, Yang X. A maximum efficiency point tracking control scheme for wireless power transfer systems using magnetic resonant coupling. IEEE Transactions on Power Electronics. 2015;**30**(7):3998-4008. DOI: 10.1109/TPEL.2014.2349534

[27] Ko WH. Early history and challenges of implantable electronics. ACM Journal on Emerging Technologies in Computing Systems (JETC). 2012;**8**(2):1-9. DOI: 10.1145/2180878.2180880

[28] Soares IV et al. Wireless powering efficiency of deep-body implantable devices. IEEE Transactions on Microwave Theory and Techniques. 2023;**71**(6):2680-2692. DOI: 10.1109/TMTT.2022.3231492

[29] Basir A, Shah IA, Yoo H. Sphere-shaped receiver coil for misalignment-resilient wireless power transfer systems for implantable devices. IEEE Transactions on Antennas and Propagation. 2022;**70**(9):8368-8378. DOI: 10.1109/TAP.2022.3161268

[30] Lamkaddem A, El Yousfi A, Posadas VG, Vargas DS. Miniaturization of wireless power transfer for implantable devices using voltage doubler rectifier. In: 2024 18th European Conference on Antennas and Propagation (EuCAP). Glasgow, United Kingdom: IEEE; 2024. pp. 1-4. DOI: 10.23919/EuCAP60739.2024.10501202

[31] Kiourti A, Christopoulou M, Koulouridis S, Nikita KS. Design of a novel miniaturized implantable pifa for biomedical telemetry. In: Lin JC, Nikita KS, editors. Wireless Mobile Communication and Healthcare. Berlin Heidelberg: Springer; 2011. pp. 127-134

[32] Lamkaddem A, Yousfi AE, González-Posadas V, Segovia-Vargas D. Miniaturized dual band implantable antenna for implanted biomedical devices. IEEE Access. 2024;**12**:15026-15036. DOI: 10.1109/ACCESS.2024.3357488

[33] Lamkaddem A, Yousfi AE, Abdalmalak KA, Posadas VG, Segovia-Vargas D. Circularly polarized miniaturized implantable antenna for leadless pacemaker devices. IEEE Transactions on Antennas and Propagation. 2022;**70**(8):6423-6432. DOI: 10.1109/TAP.2022.3161370

[34] Shah SAA, Yoo H. Radiative near-field wireless power transfer to scalp implantable biotelemetric device. IEEE Transactions on Microwave Theory and Techniques. 2020;**68**(7):2944-2953. DOI: 10.1109/TMTT.2020.2985356

[35] Zada M, Shah IA, Nasir J, Basir A, Yoo H. Empowering remote patient monitoring with a dual-band implantable rectenna system for wireless power and data transfer. IEEE Transactions on Antennas and Propagation. 2023;**71**(12):9509-9522. DOI: 10.1109/TAP.2023.3319133

[36] Cao X, Sato H, Xu K-D, Jiang W, Gong S, Chen Q. A systematic method for efficient wireless powering to implantable biomedical devices. IEEE Transactions on Antennas and Propagation. 2023;**71**(3):2745-2757. DOI: 10.1109/TAP.2023.3240005

[37] Shah IA, Zada M, Shah SAA, Basir A, Yoo H. Flexible metasurface-coupled efficient wireless power transfer system for implantable devices. IEEE Transactions on Microwave Theory and Techniques. 2024;**72**(4):2534-2547. DOI: 10.1109/TMTT.2023.3319050

[38] Li M, Khaleghi A, Hasanvand A, Narayanan RP, Balasingham I. A new design and analysis for metasurface-based near-field magnetic wireless power transfer for deep implants. IEEE Transactions on Power Electronics. 2024;**39**(5):6442-6454. DOI: 10.1109/TPEL.2024.3354394

[39] Benbuk A, Moniz-Garcia D, Gulick D, Quinones-Hinojosa A, Christen JB. A wireless battery-free implant with optical telemetry for In vivo cortical stimulation. IEEE Sensors Letters. 2024. Art no. 6005204;**8**(5):1-4. DOI: 10.1109/LSENS.2024.3387370

[40] Zhang Y et al. An 8-shaped antenna-based battery-free neural-recording system featuring 3 cm Reading range and 140 pJ/bit energy efficiency. IEEE Journal of Solid-State Circuits. 2023;**58**(11):3194-3206. DOI: 10.1109/JSSC.2023.3276174

[41] Karimi MJ, Jin M, Dehollain C, Schmid A. A wireless power conversion chain with fully on-chip automatic resonance tuning system for biomedical implants. IEEE Open Journal of Circuits and Systems. 2024;**5**:117-127. DOI: 10.1109/OJCAS.2024.3382355

[42] Kim A, Powell CR, Ziaie B. An implantable pressure sensing system with electromechanical interrogation scheme. IEEE Transactions on Biomedical Engineering. 2014;**61**(7):2209-2217. DOI: 10.1109/TBME.2014.2318023

[43] Kaim V, Singh N, Kanaujia BK, Matekovits L, Esselle KP, Rambabu K. Multi-Channel implantable cubic rectenna MIMO system with CP diversity in orthogonal space for enhanced wireless power transfer in biotelemetry. IEEE Transactions on Antennas and Propagation. 2023;**71**(1):200-214. DOI: 10.1109/TAP.2022.3222695

[44] Zhuang H, Wang W, Yan G. Omnidirectional wireless power transfer system using modified saddle-shaped coil pair for implantable capsule robots. IEEE Transactions on Power Electronics. 2023;**38**(9):11664-11672. DOI: 10.1109/TPEL.2023.3270501

[45] Saha R, Kaffash Z, Mirbozorgi SA. Multi-resonator wireless inductive power link for wearables on the 2D surface and implants in 3D space of the human body. IEEE Transactions on Biomedical Circuits and Systems. 2024;**18**(5):1024-1036. DOI: 10.1109/TBCAS.2024.3375794

[46] Garrido-Atienza A, Guardiola M, Neira LM, Romeu J, Fhager A. Movement tracking and false positive reduction method for microwave colonoscopy systems. IEEE Journal of Electromagnetics RF and Microwaves in Medicine and Biology. 2024:1-7. (Early Access). DOI: 10.1109/JERM.2024.3476964

[47] Wang J et al. A 403 MHz wireless power transfer system with tuned Split-ring loops for implantable medical devices. IEEE Transactions on Antennas and Propagation. 2022;**70**(2):1355-1366. DOI: 10.1109/TAP.2021.3111520

[48] Jia Y, Mirbozorgi SA, Zhang P, Inan OT, Li W, Ghovanloo M. A dual-band wireless power transmission system for evaluating mm-sized implants. IEEE Transactions on Biomedical Circuits and Systems. 2019;**13**(4):595-607. DOI: 10.1109/TBCAS.2019.2915649

[49] Machnoor M, Rodríguez ESG, Kosta P, Stang J, Lazzi G. Analysis and design of a 3-coil wireless power transmission system for biomedical applications. IEEE Transactions on Antennas and Propagation. 2019;**67**(8):5012-5024